GOVERNMENTS PUSH
INFANT FORMULA

Government Push Infant Formula

First published 2017

by Irene Publishing

ISBN 978-91-88061-18-8

Irene Publishing
Sparsnäs 1010, 66891 Ed, Sweden
irene.publishing@gmail.com
www.irenepublishing.com

Layout by Jørgen Johansen

Another book by George Kent from Irene Publishing:
Caring about Hunger Irene Publishing 2016

ISBN 978-91-88061-18-8

90000

9 789188 061188

GOVERNMENTS PUSH INFANT FORMULA

George Kent

University of Hawai'i
University of Sydney
Saybrook University

CONTENTS

LIST OF FIGURES

PREFACE

Feeding infants with infant formula produces worse health outcomes than breastfeeding for both the infant and the mother, in both high- and low-income countries. Many national governments fail to do what they ought to do to regulate the manufacture and sale of infant formula and other baby foods. Here, however, the focus is on cases in which governments not only fail to regulate, but also provide positive support to the formula industry. Their actions can amount to a policy of collusion between government and industry, at the expense of families. This collusion can go far beyond anything that could be regarded as a simple error.

Not only manufacturers and marketers, but also some governments push infant formula. *Push* evokes the image of drug pushers. They draw vulnerable people to substances that turn out to be harmful to the health and economic well-being of themselves and their families. The largest-scale pushes occur where national governments are actively involved in the distribution of free or subsidized infant formula on a sustained basis. The impacts on individual infants might not be obvious, but the cumulative impact on many infants over the long term can be devastating. This book shows the patterns of subsidization in Chile, Egypt, and the United States. It is not possible to assess their health and impacts directly, partly because the governments do not facilitate that sort of research. However, the abundant studies on the impacts of using formula help us understand what results should be expected.

Apart from the distribution of formula by governments of the sort examined in these three cases, there is reason for concern about the distribution of formula in various types of emergency situations (World Health Organization 2007). The over-reliance on formula in emergencies has been examined by several different agencies, and sensible recommendations have been advanced by the Interagency Working Group on Infant and Young Child Feeding in Emergencies. The Emergency Nutrition Network tracks the issues (ENN 2016). It offers an online course on Infant Feeding in Emergencies. In this context, special attention has been given to the needs of refugee children (Infant Feeding Support for Refugee Children 2016).

The distribution of formula in emergency situations usually has been carried out by governmental and nongovernmental agencies that were not fully aware of the problems associated with the distribution of free infant formula. Most seem willing to modify their practices when they learn the recommendations and the reasoning behind them. There is little evidence of willful profiteering through emergency relief by formula sellers. They do not regularly take advantage of emergencies to win commitments from the victims to their brands. The issue of formula distribution in emergency situations will not be examined here.

Some formulas are tailored to address various special medical needs. The focus in this book is on the ordinary formula used for the general population of children under one year of age. There are serious issues relating to special formulas, but they are not reviewed in this book.

Overall, the argument here is that large-scale subsidies of infant formula should be phased out. Much more could be done to support breastfeeding, at the individual level and at the societal level. For cases in which mothers do not breastfeed, it is often possible to provide human milk from other women.

The analysis unfolds as follows.

Chapter One, *Formula is Not Second Best*, makes the point that feeding infants with formula rather than breastfeeding regularly leads to worse health outcomes for them and their mothers. Feeding with formula may sometimes be unavoidable because of other considerations, such as women's not getting the support they need at their workplaces. The appropriate remedy then is to provide the support. If an alternative to breastfeeding by the mother is needed, it may be feasible to provide human milk from other women.

Chapter Two, *Chile*, assesses the infant formula distribution program in Chile, based on a study prepared in collaboration with a Chilean expert on infant feeding, Cecilia Castillo (Kent and Castillo 2016).

Chapter Three, *Egypt*, assesses the formula distribution program in Egypt, based on a study prepared in collaboration with two Egyptian experts on infant feeding, Ghada Sayed and Azza Abul-Fadl (Kent, Sayed, and Abul-Fadl 2017).

Chapter Four, *United States*, was prepared despite resistance to its line of analysis from the California WIC Association. I was invited to give a talk to their annual conference in 2011. After having been listed in the program and having arranged flights and hotel bookings, they understood that I would have some critical things to say. They suggested specific talking points of praise for WIC's activities that I might use instead, but I declined. I was disinvited just days before the event.

Chapter Five, *The Marketing Code*, examines the relevance of the International Code of Marketing of Breast-milk Substitutes. It is placed near the end of the book to make it clear that the core concern here is the harm to the well-being of families, and not just the legal technicalities in the violation of an international document. The central theme of the book is that national governments sometimes act in collusion with formula makers and sellers who profit at the expense of families' well-being.

Chapter Six, *Infant Feeding and Human Rights*, places this discussion into the global human rights framework, as it is and as it could be. It is based on my frequent writing on children's right to food, beginning with my book on *Children in the International Political Economy* (Kent 1995).

Chapter Seven, *Human Milk*, explores what could become a practical innovation. It suggests that well-managed and well-regulated human milk banks and systems of milk sharing could make infant formula almost obsolete, to be used only in cases of clear medical needs, like medicine

Chapter Eight, *The Inadequacy of Infant Formula*, argues that the claims of benefits from using specific infant formulas and various additives have been misleading, not just occasionally, but systematically. The failure of governments to challenge manufacturers' claims is another component of the collusion between governments and the formula industry. Governments have an obligation to protect, promote, and support the interests of infants and their families.

This book is being published just months after my book, *Caring About Hunger*, also by Irene Publishing. The link between them is simple and direct. When governments act in ways that benefit industry, and do that at the expense of young children, it reveals what they care about. This is a case study about caring.

ACKNOWLEDGMENTS

This book project emerged out of decades of work on food issues from a politics and policy perspective. I have been supported along the way by many wonderful people. For their help in dealing specifically with infant feeding issues, I want to thank James Akré, Azza Abul-Fadl, Phillip Baker, Carol Bartle, Mike Brady, Adriano Cattaneo, Cecilia Castillo, J.P. Dadhich, Sharon Friel, Ted Greiner, Arun Gupta, Elisabet Helsing, Alessandro Iellemo, Urban Jonsson, Michael Latham, Lida Lhotska, Alison Linnecar, Maureen Minchin, Valerie McClain, Mary Renfrew, Libby Salmon, Ghada, Sayed, Julie Smith, Betty Sterken, Pamela Morrison, and Patti Rundall.

I also want to give special thanks to the co-authors of articles on Chile and Egypt that provided the foundations for Chapters Two and Three, already acknowledged in the Preface, and to Maureen Minchin for her insightful and detailed critical assessment of the manuscript.

I am grateful to the photographers and artists who produced the images used in this book. I particularly want to thank Professor Emeritus Stephen Buescher for allowing me to use the image on the cover and Figures 8-1 and 8-2. Figure 4-1, of unknown origin, was found on the Internet. Figure 4-2 is from the Etsy website provided under the image. The photograph on the back cover, by Mallory Smothers, is used with her permission.

Finally, I want to express my gratitude for the unwavering support of my family, Joan, our sons, Greg and Jeff, and their wives, Kelli and Tina. Among their many accomplishments is their providing us two wonderful grandchildren, Keiko and Alice. All of them helped propel this project in their own ways.

CHAPTER ONE

FORMULA IS NOT SECOND BEST

Infants should be breastfed (UNICEF and WHO 2015). The World Health Organization's *Global Strategy for Infant and Young Child Feeding* is clear about the importance of breastfeeding:

> The vast majority of mothers can and should breastfeed, just as the vast majority of infants can and should be breastfed. Only under exceptional circumstances can a mother's milk be considered unsuitable for her infant. For those few health situations where infants cannot, or should not, be breastfed, the choice of the best alternative – expressed breast milk from an infant's own mother, breast milk from a healthy wet-nurse or a human-milk bank, or a breast-milk substitute fed with a cup, which is a safer method than a feeding bottle and teat – depends on individual circumstances. (World Health Organization 2003, 10)

At the global level, the Codex Alimentarius Commission develops non-binding guidelines regarding food composition and safety, including infant formula (Codex Alimentarius Commission 2016). The Food and Agriculture Organization of the United Nations and the World Health Organization jointly established the Commission in 1963. Its main purposes are "protecting health of the consumers and ensuring fair trade practices in the food trade, and promoting coordination of all food standards work undertaken by international governmental and non-governmental organizations." It does this by developing food standards, guidelines and related texts such as codes of practice under the Joint FAO/WHO Food Standards Program (Codex Alimentarius Commission 2011).

In 1976, at its 11th session, the Codex Alimentarius Commission issued a *Statement on Infant Feeding*. It said, "it is necessary to encourage breastfeeding by all possible means in order to prevent that the decline in breastfeeding, which seems to be actually occurring, does not lead to artificial methods of infant feeding which could be inadequate or could have an adverse effect on the health of the infant (Codex Alimentarius Commission 1976)."

At this session in 1976 the Commission adopted a Codex Standard for Infant Formula. The standard, designated as CODEX STAN 72-1981, includes a list of required ingredients and names various required quality control measures. In 1983, the 15th Session adopted amendments to the sections on Food Additives and Labeling. A further amendment to the Labeling section was adopted in 1985 by the 16th Session. Amendments to the Vitamin D and B12 amounts were adopted by the 17th (1987) and 22nd (1997) sessions (Codex Alimentarius Commission 2007).

This core statement of the required ingredients for infant formula was based on the formula products in use at that time. The permitted nutrient ranges legitimized a variety of very different formulas. The Codex requirements are widely regarded as a minimum standard. Some countries have adopted more stringent requirements. Many important additions have been made to infant formula recipes to remedy deficiencies and excesses recognized over time.

Most infant formula is made from powdered cow's milk, augmented with required ingredients and some optional ingredients. In place of cow's milk, some formulas are based on soy milk or goat's milk.

Chapter Eight provides additional discussion on the composition of infant formula.

It is widely recognized, even by infant formula manufacturers, that feeding with formula regularly leads to worse health outcomes than breastfeeding for both mother and child. Many studies have demonstrated this (Bartick and Reinhold 2010; Bartick et al. 2016; Chen and Rogan 2004; Ip et al. 2007; Piwoz and Huffman 2015; Payne and Quigley 2016; Rollins et al. 2016; Stevens, Patrick, and Pickler 2009; The Lancet 2016; Thurow 2016; World Health Organization 2013a, 2013b; Zimmerman 2016). The only possible conclusion is that infant formula is inadequate (Kent 2012a, 2014b), a theme examined in detail in Chapter Eight.

The studies just cited may underestimate the damage done by formula feeding. Maureen Minchin argues that formula has been responsible for cumulative intergenerational damage across whole populations (Minchin 2015b). She sees formula as the single most important postnatal factor in modern epidemics of inflammatory disease, and offers substantial evidence to support that view.

Improved infant feeding practices could do a lot to reduce child mortality and morbidity worldwide. The dimensions of the need can be estimated by looking at child mortality data.

In 2013, 6.3 million children died before their fifth birthdays (World Health Organization 2014). To get a sense of scale on this, this is far more than the number of deaths due to armed conflict in that year. In 2013 the twenty deadliest wars in the world resulted in an estimated 127,134 deaths (PS21 2015; World Bank 2015). The world's mass media gave far more attention to the thousands of deaths in armed conflict than to the millions of deaths of children.

About 45% of all child deaths are linked to malnutrition (World Health Organization 2014). "Globally, breastfeeding has the potential to prevent about 800,000 deaths among children under five each year if all children 0–23 months were optimally breastfed (World Health Organization 2015; also see Sankar et al., 2015)."

The risk of dying declines steadily as children get older. The highest rate of infant mortality occurs during the neonatal period, the first 28 days of life. This is partly due to problems *in utero*, before the infant is born, such as problems associated with malnutrition or illness of the mother during her pregnancy. Many of those neonatal deaths are associated with delayed or absent breastfeeding. One study says, "16% of neonatal deaths could be saved if all infants were breastfed from day 1 and 22% if breastfeeding started within the first hour (Edmond et al. 2006, e380)." Where for some reason early direct breastfeeding by the mother is not possible, the use of banked human milk can help to reduce neonatal deaths. This would apply not only to critically ill infants but to all infants, especially preterm and small-for-gestational-age infants.

Child mortality data describe only part of the harms associated with sub-optimal infant feeding practices. The increased mortality risk continues beyond the child's fifth birthday. Poor practices also increase morbidity, especially in terms of higher infection rates and worse physical, cognitive, and intellectual development. Some harms continue on into adulthood (Grummer-Strawn and Rollins 2015; World Health Organization 2013a; 2013b).

Apart from the health risks, feeding with formula can result in substantial economic harm because of the costs of formula and other processed foods that are likely to follow, and also because of increased health care costs.

In some countries, especially in southern Africa, fears about HIV transmission through breastfeeding have led to large-scale provision of free formula, sometimes to women who had not been diagnosed as HIV-positive. The risks of transmission are now sharply reduced with appropriate drug treatments, but some governments find it difficult to reduce the supply of free formula. Governments that provide free or subsidized infant formula can find it difficult to back away from that policy when they get new information or conditions change. In some cases the option of providing stronger support for breastfeeding seems to be forgotten, and instead the supply of subsidized formula grows.

In Chile, the government has been providing fortified cow's milk for a long time, so infant formula seemed an improvement.

In Argentina, the city of Buenos Aires has agreed to a new arrangement with Nutricia Bagó (an affiliate of Danone, based in France), to train people in charge of child care centers how to feed young children. The movement is raising alarm among breastfeeding advocates in the country. The journalist who investigated this asked, "with breastfeeding in slow recovery, will it be the governments who ruin it again (Barruti 2016a, 2016b)?"

Argentina's province of Córdoba started giving one can per month of Nestlé's starter formula, Nidina, as a nutritional complementary food for families under the poverty line. There is no indication of any need for it. The program does not include any sort of breastfeeding support.

In Saudi Arabia, there have been many concerns. In 2014 the suppliers of infant formula were cautioned about their increasing retail infant formula prices (Astley 2014). In 2015, infant formula subsidies shot up sharply (AmeInfo 2015). In 2016 the cabinet decided to have the Ministry of Health provide milk powder free of charge for infants up to two years of age in poor families (Life in Saudi Arabia 2016; Saudi Gazette 2016; Toumi 2016).

These governments' intentions may be good, but one must wonder who is advising them. The late James Grant, formerly Executive Director of the United Nations Children's fund, made it clear that supporting breastfeeding is the best thing governments could do for infants in poor families:

> Breastfeeding is a natural "safety net" against the worst effects of poverty. If the child survives the first month of life, the most dangerous period of children, then for the next four months or so, exclusive breastfeeding goes a long way toward cancelling out the health difference between being born into poverty and being born into affluence . . .
>
> It is almost as if breastfeeding takes the infant out of poverty for those first few months in order to give the child a fairer start in life and compensate for the injustice of the world into which it was born. (CEPPs 2016)

Distribution of infant formula by governments is problematic because the provision of free or subsidized formula, together with the apparent government endorsement of the product, tends to increase formula consumption and displace breastfeeding, resulting in harm to the health of both infants and their mothers.

Good intentions of governments can be subverted by the lobbying efforts of the formula industry. In 1974, there were fears that the United States Congress might not fund the new WIC program (see Chapter Four):

> Apparently the companies' market research led them to despair of ever convincing the low-income groups, especially African Americans, to switch from evaporated milk to infant formula. This gave them the ingenious

> "health" argument that offering free formula was the
> only way to get low-income U.S. populations weaned
> off of such a relatively less appropriate means of feeding
> infants (with, for example, accompanying high levels of
> infant anemia). (Greiner 2011)

Formula originally got the government's attention because
it was better than the evaporated milk. The alternative of
providing stronger support for breastfeeding was not given much
consideration. Formula became the key element in the government's
food supplementation program.

Many families get free or subsidized formula even though there is
no good medical or economic reason for it. Many want formula
because the government seems to endorse its use. In most cases,
there would be more benefits from providing strong support for
breastfeeding, at the individual and societal levels, or providing
human milk from well-regulated and well-managed milk banks.

Many people think of formula as a good second-best to be used
when the biological mother does not breastfeed, but the first
option should be more vigorous support for breastfeeding for
the individual mother and in the society as a whole (Lee 2016).
Direct advice from skilled lactation counselors can be very helpful.
Better maternity support through programs such as extended paid
leave are important, as are accommodations for breastfeeding at
the workplace and in public places such as railway stations and
airports. When conventional direct breastfeeding cannot be done,
the mother may be able to express her milk and place it into bottles
or pouches,, and have it delivered to the infant by herself or another
caretaker.

As discussed in Chapter Seven, if enough good quality breast-milk
cannot be obtained from the biological mother, it is human milk

from other women that is next best. Milk sharing arrangements or human milk banks can be developed under appropriate guidelines and regulations. The human milk they provide would be better for infants' health than any manufactured formula.

It is not only breastmilk, but skilled support for breastfeeding that is needed.

> UNICEF analyses show that women are not getting the help they need to start breastfeeding immediately after birth even when a doctor, nurse of midwife is assisting their delivery. In the Middle East, North Africa and in South Asia, for example, women who deliver with a skilled birth attendant are less likely to initiate breastfeeding in the first hour of life, compared to women who deliver with unskilled attendants or relatives. (UNICEF 2016b)

Globally, women who deliver with a birth attendant are *less* likely to meet that requirement for optimum breastfeeding. In some cases, this might be because marketers of infant formula promote their wares through birth attendants, convincing them that formula should be used right from the beginning of life. Much too often, infant formula is provided to newborns in hospitals without the consent of the parents, and without any good medical justification.

Without initiation of breastfeeding during the first hour after birth, infants are not likely to get the first milk, the colostrum and the milk that follows it, so critically important in the development of the infant's immune system. Beyond that, the provision of formula in the first hours and days of the newborn's life can undermine the potential for establishing a strong breastfeeding relationship between mother and child (Boban and Zakarija-Grković 2016 Handley and Meta 2016). This has been one of the major concerns driving the global Baby Friendly Hospital Initiative (UNICEF

2016c). Agencies that oversee birthing sites should fully recognize the importance of early initiation and continued exclusivity of breastfeeding.

Feeding with infant formula displaces breastfeeding in two ways. In the short term, an infant who gets more of one will get less of the other. In the long-term, when a young infant is fed with formula instead of its mother's breastmilk, the mother's ability to produce breastmilk diminishes. If the infant and mother deviate from the breastfeeding path, it becomes difficult to return to it. The distribution of free or subsidized infant formula actively suppresses breastfeeding, with lasting effects.

The resources now devoted to subsidizing formula could instead be used to better protect, promote, and support breastfeeding. Government policy should be attuned to current global recommendations (UNICEF 2016a, 2016e). If governments spent less on formula and more on providing well-trained birth attendants and other support services, health outcomes for infants and mothers would be much better.

The family itself is the primary support system. Mothers and fathers need support so they in turn can provide the best possible support for their new child, both before and after the birth event. Many governments require paid leave not only for the mother but also for the father. The United States is the only high-income country that does not require paid leave for either of them. Some countries provide several years leave for the mother, with guarantees of job security (OECD 2016).

MARKET SHARE

Choosing between alternative methods of feeding infants might seem a simple private family-level issue, but when viewed at the national and global levels, we see an intense economic and political struggle for market share. What proportion of the world's infants' food intake should come from their mothers and what proportion should come from the formula industry

The factories are winning the struggle. There is a huge effort underway to promote the use of infant formula worldwide, especially in emerging economies with a growing middle class. Globally, the infant formula industry is reported to be growing at more than 11% a year. There is increasing alarm about the vigorous promotion of infant formula and the impact it is likely to have on the health of infants and the adults they will become (Baker et al. 2016; Kent 2006b, 2011, 2015; Lisa 2015; Smith, Salmon, and Baker 2016).

Most of the criticism has focused on the manufacturers and marketers, in articles with titles such as "Formula for Profit: How Marketing Breastmilk Substitutes Undermines the Health of Babies (Coburn 2000)." However, some critics pay attention to the role of government, under titles such as "Formula for Profit: How Private Corporations Grow Fat from a Program Designed to Feed the Poor (Novac 1988)." Judith Richter has been one of the few to call governments to account for their failure to hold the formula companies accountable for their actions that put infants at risk (Richter 2001).

In the struggle for market share, the balance has been tipped in favor of expanded formula distribution because of the economic and political influence of the formula manufacturers and sellers, often working together with national governments. Their push is driving a deterioration of breastfeeding practices in many parts of the world.

Government policy is an important factor in this struggle. Regulatory practices can favor either infants or commerce. Similarly, subsidy programs can favor either infants or commerce. I examined the regulatory mechanisms, particularly the United States Food and Drug Administration, in a previous book (Kent 2011). This book examines large-scale government programs for subsidizing infant formula, a practice that should get much more critical attention.

CHAPTER TWO

CHILE

THE PNAC PROGRAM

The widespread distribution of free milk products for children by national governments can be traced back at least to France's initiative in 1900. Since then, the practice has been adopted by many countries. Chile's program, Gota de Leche (A Drop of Milk) was launched in San Bernardo in 1911 and quickly spread to many other parts of the country (Memoria Chilena 2016.

The Programa Nacional de Alimentación Complementaria (PNAC) was established in 1987. Its title is commonly translated as the National Complementary Food Program, but it is better described as a supplementary food program because it is intended to supplement diets that might be deficient. In other contexts, complementary foods are those provided to infants to complement breastfeeding beginning around six months of age.

The program addresses the nutritional needs of children under six years of age and their mothers. The service is divided into several categories:

> **PNAC Básico**, for healthy mothers and infants. Purita Fortificada can be obtained for infants up to 18 months of age.

PNAC Refuerzo, for infants who are malnourished or at risk of malnutrition
PNAC Prematuros, for infants who are born prematurely. They are given formula specially designed for premature or low-birth-weight infants.
PNAC de Enfermedades Especiales, for children with special diseases. In this category, some benefits may be continued until the child reaches 18 years of age.

PNAC Básico serves the largest number of infants. It provides a variety of cow's milk-based food products for both children and their mothers under various rules. The composition of the foods provided and the rules have varied over time, based principally on the emergence of new understandings about nutritional needs.

Infants are entitled to Leche Purita Fortificada, a powdered milk product fortified with vitamin C, iron, zinc, and copper. The fortified product replaced the plain cow's milk with no additives that PNAC had distributed before 1998 (Rama de Nutrición 1999). It comes in a one-kilogram package. The Ministry of Health occasionally describes it as a type of formula because they add ingredients to the cow's milk, but it is not designed to meet widely accepted international standards for infant formula (Codex Alimentarius Commission 2007).

PNAC advises mothers who provide Purita Fortificada to infants under one-year-old to dilute it and add vegetable oil and sugar, or preferably, maltodextrin. These additives are not provided by PNAC, but must be purchased by the mothers in local stores.

The amounts of Purita Fortificada to which families are entitled are based on their feeding patterns. Women who **exclusively breastfeed** (Lactancia Materna Exclusiva) during the first six months get none during those six months or afterward. Internationally, exclusive breastfeeding is understood to mean no water or formula of any kind, but for the purposes of this program "exclusive" is taken to mean at least 90 percent of the daily feeds consist of breastfeeding.

Women who **predominantly breastfeed** (Lactancia Materna Predominante) get one kilogram a month during the first two months, and then two kilograms a month up to and including the fifth month, and none after that. Women who predominantly breastfeed are those who breastfeed between 50 and 90 percent of the time.

Women who feed **formula predominantly** get two kilograms during each of the first two months, and three kilograms during the third through fifth months, and no formula after that. They also get three kilograms of a cream soup mixture during the fifth month.

Those who feed **formula exclusively** get two kilograms during each of the first two months and three kilograms during the third through eleventh months. They get two kilograms from the twelfth through the seventy first months—up to the child's fifth birthday. They also get one kilogram of a cream soup mixture during the fifth month and two kilograms of the cream soup from the sixth through the seventy-first months (Ministerio de Salud 2011, 10-11).

These arrangements are changing because of a new initiative to replace at least some of the Purita Fortificada with starter formula, *fórmula de inicio* (Ministerio de Salud 2016a, 2016b, 2016c 2016d). The plan document reviews current knowledge of the importance of breastfeeding for both the infant and the mother's health and calls for vigorous support when there are difficulties in breastfeeding. Pilot studies on the distribution of a starter formula were started in 2016. The prevailing view in Chile is that starter formula is to be used for only the first few months of the infant's life. (In other places, starter formula is recommended to be used for twelve months.) It is not yet clear what would replace Purita Fortificada as the children get older and starter formula is no longer provided.

The plan identifies conditions of the mother or child that would make them eligible for free formula from PNAC. If those conditions are not met, there is a possibility of receiving the formula, but only if the mother signs an informed consent form and attends a counseling session on breastfeeding.

CRITICAL ANALYSIS

The *Proyecto de Incorporación de Fórmula de Inicio en el Programa Nacional de Alimentación Complementaria (PNAC)* (Ministerio de Salud 2016a) recognizes that, in terms of its impact on infants' health, Purita Fortificada is inferior to the starter formulas offered by major manufacturers. However, both tend to displace breastfeeding. The central concern here is that the new starter formula distribution program might expand to a much larger scale, with more infants consuming it, and with each infant consuming more formula. If it does go to a larger scale, there will be much less breastfeeding. This issue deserves close attention because of the impact it could have on the health of Chile's children and the adults they will become.

Chile's plan for distributing starter formula says that if support for breastfeeding fails, the second best alternative is feeding with formula. As the preceding chapter explained, that is not correct.

The plan says that formulas made from cow's milk have been improved in such a way as to approach the quality of breastmilk (Ministerio de Salud, 2016a, 3). The source cited to support their point (Ministerio de Salud 2014) does not claim that formula is close to the quality of breastmilk.

In discussing the addition of fatty acids to infant formula, the plan says the benefits have been well demonstrated (Ministerio de Salud, 2016a, 7). That is not correct (Kent 2014a; Minchin 2016).

The pilot studies involve both changes in the product that is supplied (Purita Fortificada versus starter formula) and the rules under which they are supplied. It will be difficult to distinguish which changes in impacts are due to changes in the product and which are due to changes in the rules. It would be wise to conduct studies in which only the rules are changed. Rules that limit the availability of free food for infants would be likely to increase the breastfeeding rate, and thus improve infants' health, regardless of which food is offered.

Table 1 of the planning document (Ministerio de Salud 2016a) lists conditions under which women can receive starter formula from PNAC for their infants. For example, they include diagnoses with HIV/AIDS or herpes, and mothers who use drugs incompatible with breastfeeding or who are undergoing certain types of chemotherapy. This list of conditions was based on the World Health Organization's list of medical reasons for using breast-milk substitutes (World Health Organization 2009).

Much of that information is now outdated. For example, instead of being advised to use formula, women with herpes lesions on their nipples could be supported in expressing their breastmilk and feeding it to the infant with a tube or bottle.

31

Women who are diagnosed as HIV-positive are not always advised to avoid breastfeeding because of fear of transmitting the virus to their infants through the breastmilk. Antiretroviral drugs can be taken by breastfeeding women to sharply reduce the risk of HIV transmission while preserving the health advantages of breastfeeding (American Academy of Pediatrics. Committee on Pediatric AIDS. 2013; Langa 2010; National Institutes of Health 2016; World Health Organization and UNICEF 2016). For a time, the United Nations Children's fund distributed infant formula to prevent HIV transmission through breastfeeding, but it has discontinued this practice because many more children died from being fed with formula than from HIV infection (de Wagt and Clark 2004).

The WHO document on medical reasons to use breastmilk substitutes (World Health Organization 2009) did not anticipate the rapid advances in the establishment of human milk banks in recent years. Worldwide, many critically ill infants in neonatal intensive care units who are too weak to suckle are provided with their own mother's milk indirectly, or other women's milk, through human milk banks. (See Chapter Seven.)

It is unfortunate that Chile has only one human milk bank for the entire country (Viñals 2015). Brazil has more than 200 (Fox News Latino 2014). If Chile had a fully developed network of milk banks, there would be less need for infant formula.

Chile anticipates providing formula if mothers or infants have the specific conditions identified in the WHO document on medical reasons for not breastfeeding (World Health Organization 2009). In addition, some mothers would be able to obtain formula if they sign informed consent statements. The statement would say that the woman is aware of the benefits of breastfeeding and, having attended a counseling session about it, she nevertheless wants to forego breastfeeding (Ministerio de Salud 2016d, 5). It is not clear whether there would be any limits to the number of families that could use this consent procedure and get free formula on request. Chile could end up following the pattern in the United States, described in Chapter Five, where the government provides free formula for about half the infants in the country.

Chile's plan makes a brief reference to the human right to adequate food, and points out that under that body of law the state cannot deny children's right to adequate food (Ministerio de Salud 2016d, 5). However, "cannot deny" does not mean the state is obligated to provide that food (Kent 2005).

The World Health Organization document was about medical reasons for using breast-milk substitutes. It was not about the conditions under which free formula should be supplied by the government.

Chile's plans for explaining the risks of formula feeding to new parents are not clear. Parents need good information if they are to make well-informed choices about how to feed their infant.

Policymakers also need good information to make well-informed choices. The government's plan for providing starter formula should be based on a thorough analysis of all significant impacts for the government, the families, and the environment, short term and long term.

Purchasing the formula would be very costly for the government, as seen already in the pilot program. The costs for staff and offices to operate the program also would be substantial, especially where there are complex criteria and rules for deciding who is to get what.

The program would be costly to families for several reasons. The free formula would be provided only for a limited time, after which the families would have to pay for formula themselves. Mothers who breastfeed for the first six months would be able to continue breastfeeding, along with providing other foods, after that first six months. Women who stopped lactating would have to purchase the formula they wanted.

In Chile, less educated mothers are more likely to breastfeed (Ministerio de Salud 2013). Formula is highly valued, and apparently endorsed by government. The government's making formula available for free attracts mothers to it and reduces breastfeeding rates. It could also lead to greater dependence on commercial processed foods over the long term, thus increasing these families' economic burdens.

The government could consider using generic formula rather than well-known brands. However, the producers of generic formula would be less likely to bid to supply the product. As illustrated by the pattern in the United States, the makers of generic formula would be reluctant to bid because they would not be able to reap the benefits of winning brand loyalty. That brand loyalty would be based on the apparent endorsement by the government of whatever brand wins the contract.

ETHICAL DILEMMAS

Chile's Ministry of Health has been concerned about what it sees as an ethical dilemma. People with low incomes get Purita Fortificada from the government for free while people who have money can purchase what they believe to be higher quality commercial infant formula in the marketplace (Stipicic 2016). This could be a serious matter. Offering starter formula would result in an increased incentive to seek formula from PNAC and a reduced incentive to breastfeed.

There are other ways to respond to that dilemma. Support for breastfeeding could be strengthened without introducing a new formula product. Well-funded lactation clinics could be established in every health center. Some of the money that governments would devote to buying infant formula could instead be used to start well-managed human milk banks, or to provide general food subsidies for people with very low incomes. Human milk sharing programs could be developed. Many things could be done without subsidizing a new type of formula.

The supply of free food by the government in Chile began when the country was poor and malnutrition was a serious problem. Now that conditions have greatly improved the rationale for this support for the general population is weak. It still makes sense to supplement the diets of people who are very poor or have special needs, but in modern times that should be done only for a small portion of the population. If free food is provided, it should not be an expensive highly processed food that is known to have negative impacts on health. There is no equity argument that justifies the distribution of free infant formula.

What is the problem that would be solved by distributing free infant formula? Clearly, there is a need for serious discussion of the alternatives.

The implementation proposal for the project calls for defining a target group for the new product, and for strengthening actions to protect breastfeeding in that target group (Ministerio de Salud 2016a, 3). However, the title of the document emphasizes that the project is about introducing starter formula into the PNAC program. It looks like the primary purpose of the project is to introduce starter formula and the secondary purpose is to promote breastfeeding.

The Ministry of Health said it intends to work toward increasing the prevalence of breastfeeding. If that is the most important objective, the priorities should be reversed. The entire project would be more attractive if the first priority was to increase breastfeeding rates.

The government could make an explicit commitment to *steadily reduce the number of infants getting either Purita Fortificada or starter formula from the PNAC program.* Families would still be able to purchase these products in the markets in the same way they purchase other foods. This phase-out would only mean that the government would not provide the products. The government of Chile could instead provide stronger support for breastfeeding. For women who do not breastfeed, the government could make banked or shared human milk readily available.

CHAPTER THREE

EGYPT

Egypt has a long history of providing subsidized food for its people. The policy has been criticized for its high cost to the government and for the way in which sustained provision of food by the government can be disempowering for the people (Ecker, Tan, and Al-Riffai 2014; IFPRI 2013). A recent study of the country's food system did not discuss the subsidies for infant formula, apparently because infant formula was not part of the national food subsidy system managed by the Ministry of Internal Trade (Ecker et al. 2016). This is unfortunate, especially in view of the study's description of the widespread malnutrition among Egypt's infants and young children.

This chapter focuses on the provision of subsidized infant formula by Egypt's government. It is important because feeding with formula displaces breastfeeding, resulting in worse health outcomes for both infants and mothers. The use of infant formula should not be encouraged. There are few conditions under which it is medically necessary.

HISTORY

Subsidized infant formula has been made available through the government's Egyptian Pharmaceutical Trading Company for decades (Egyptian Pharmaceutical Trading Company 2016). Since no formula is manufactured in Egypt, the company has been importing formula and providing it to pharmacies and Primary Health Care Clinics throughout the country.

The prices paid by the government and by families for infant formula have been increasing. One report suggests that companies importing formula created a monopoly to increase prices (Ahram Online 2016). There has been serious concern about leakage of infant formula from the supply chain, largely due to corruption. Some ends up being used by infants who do not need subsidized formula. Some is diverted to illegitimate uses such as making cakes and biscuits in bake shops.

These factors led to serious shortages in the supply of subsidized formula. There have been allegations that some subsidized formula has instead been sold at a higher price, thus making the shortage of subsidized formula especially intense. According to one report, "rampant corruption was discovered with some health employees and pharmacists making huge profits by selling the EGP5-subsidised formula at as high a price as EGP60 (Sadeq-Nevine and Sidqy 2016)."

Because of concerns about the way formula distribution was managed by the Egyptian Pharmaceutical Trading Company, in June 2016 the Minister of Health and Population (MoHP) ordered radical changes in the system. To reduce leakage, a Smart Card system was introduced to identify those who were eligible, and the criteria for eligibility were modified.

Over the summer of 2016, national policy in relation to infant formula changed rapidly. The MoHP ordered that all formula would be distributed through the pharmacies of the Primary

Health Care Units throughout the country. The new approach was described as follows:

> Starting on August 1, 2016, beneficiaries eligible for the fully subsidized infant powdered milk formula, the red can, must use their smart cards to get their allowances. According to MoHP, beneficiaries have to meet the following conditions: the mother is not capable of breastfeeding her infant, passed away, delivered more than one child, has a disease(s) that prevents her from breastfeeding, or takes medicines [that] can negatively affect the child through her milk.
> In order to obtain the smart card, beneficiaries must apply at any of the Egyptian Postal Service's branches with the supporting documents and MoHP's approval. Beneficiaries will receive their monthly allowance using their smart cards at MoHP's Maternal and Child Centers that are equipped with card readers. MoHP stated that it has around 1005 centers in all governorates equipped with the card reading machines. (United States Department of Agriculture 2016i)

Steep price increases and shortages, together with inadequate information to the public about the changes, led to street protests in early September 2016. Egypt's Armed Forces were given control of the infant formula program (Ahram Online 2016; Aswat Masriya 2016a, 2016b, Middle East Monitor 2016; Sadeq-Nevine and Sidqy 2016; Saleh 2016; Tablawy 2016). All these difficulties raise important issues about the management of the system of providing subsidized infant formula. However, the deeper long-term questions are about the wisdom of providing such subsidies at all.

CRITICAL ANALYSIS

In early September 2016, the Ministry of Health and Population changed the criteria under which families would be eligible for subsidized formula several times. Some criteria were clear, such as the birth of twins or triplets or the death of the mother. Some criteria were not so clear. A family would be eligible if the woman stopped breastfeeding, but no clear distinction was made between being unable to breastfeed and choosing not to breastfeed. If any woman could get subsidized formula simply by choosing not to breastfeed her infant, there would be no need for additional detailed criteria and examination procedures. The criterion for eligibility could be reduced to "not breastfeeding". That would be equivalent to having no criteria at all, and would amount to offering subsidized formula on request.

Maternal diabetes was said to warrant the provision of subsidized formula, but the reason for that was not explained. It was stated that chronic illnesses "such as" diabetes would justify claims for subsidized formula, but the meaning of "such as" was not explained.

The use of diabetes as an example is ironic as diabetes is more common in children who are exposed to cow's milk formula. Breastfeeding improves diabetic control for the mother, reducing her insulin needs (Minchin 2015b).

The criteria did not explain how "weakness in milk production" would be addressed. Apparently, the plan was that doctors or nurses would examine women's breasts but many people felt this would violate women's dignity and had no sound scientific basis (El-Bar, Karim, 2016; Mahfouz 2016; Wirtschafter 2016). Both the media and the people generally seemed unaware of the existence of professional lactation counselors whose primary role is to help new mothers deal with difficulties in breastfeeding.

Poverty was not specified as a criterion of eligibility for subsidized formula.

Viewed over the long term, Egypt's entire food subsidy program should be questioned. While it was intended to increase food security, especially for the poor, it may have done the opposite, leading the poor to become overly dependent on unreliable food supplies from the government.

Here, however, the focus is on the infant formula program. The logistical issues raised in mid-2016 are important, but it is more important to reconsider the wisdom of providing subsidized infant formula over the long term.

From its beginning, Egypt's program for distributing subsidized infant formula has been based on the premise that if an infant cannot be breastfed by its mother, the second best alternative is feeding with formula. That is not correct, as explained in Chapter One.

In Muslim countries such as Egypt there are religious issues relating to the use of human milk obtained from a woman other than the biological mother, even though, as an infant, the Prophet Mohammed was fed by a wet nurse. Banked or shared human milk might be acceptable under some conditions.

Traditionally, milk-kinship has been understood by Muslims as meaning that infants breastfed by the same woman must be viewed as siblings, and therefore must never marry. However, there was a new fatwa in 2004 "stating that making use of such milk banks, in case of need, does not raise religious problems in Islam, adding that using such milk does not institute milk-kinship which prohibits marriage in Islam (Ghaly 2010, 5)."

41

The issues should be explored further to identify conditions under which the use of banked or shared human milk might be compatible with Islam (al-Naqeeb, Azab, and Mohammed 2000; Khalil 2016; Thorley 2016).

The World Health Organization and many other global and national agencies recommend initiation of breastfeeding within the first hour after birth with direct skin-to-skin contact, exclusive breastfeeding for the first six months, and continued breastfeeding along with other foods for up to two years and beyond. These conditions for optimal infant feeding frequently are not met in Egypt.

Surprisingly, "Both medical assistance at delivery and delivery at a health facility are associated with lower proportions of children for whom breastfeeding was initiated within the first day of birth . . . (El-Zanaty and Way 2009, 165)." Why would such care lead to worse breastfeeding practices? If medical assistance and delivery at a health facility do not lead to better feeding practices, the quality of care must be questioned.

Regarding exclusive breastfeeding for the first months of life, it is important to consider the trend over time. The EDHS survey for 2000 said, "About one-third of children are exclusively breastfed throughout the first 4-6 months of life (El-Zanaty and Way 2001, xxv)." The 2008 survey said 28.8 percent were exclusively breastfed at 4-5 months (El-Zanaty and Way 2009, 168). The survey for 2014 found, only 13 percent are exclusively breastfed (EDHS 2014, 26-27).

When compared with internationally recommended infant feeding practices, Egypt has been moving in the wrong direction. It seems clear that the decline in exclusive breastfeeding has been due to the combination of inadequate support for breastfeeding and the subsidization of infant formula by the government.

In early September 2016, it was reported that "the infant milk subsidy eats up about EGP450 million of the State Budget (Sadeq-Nevine and Sidqy 2016)." On September 6, 2016, the Egyptian Medical Syndicate called on the Ministry of Health and Population and the Ministry of Finance to ensure that the budget for subsidizing infant formula must be not less than twice that amount, 900 million Egyptian pounds (about 50 million U.S. dollars), and it insisted there should be no reduction in the subsidization of infant formula (Al Ghad 2016). It is not clear why Egypt's physicians would take this position. There is no compelling medical reason for the large-scale sustained subsidy of infant formula by any government. It puts at risk the health of both infants and their mothers.

This support for formula might be explained by the fact that in Egypt, "Artificial baby food companies spend millions of pounds on buying the goodwill of physicians by way of gifts, participation in conferences, etc. (IBFAN 2011)." The companies follow this practice in many countries.

In Egypt, subsidizing infant formula harms infants' health, leads to corruption and political turmoil, places a burden on the country's foreign exchange resources, and generally causes more problems than it solves. It should be phased out for both health and economic reasons.

The resources now being devoted to a poorly functioning system for subsidizing formula could instead be used to provide better protection, support, and promotion for breastfeeding. In designing that support, government policy should stay attuned to current global recommendations (UNICEF 2016).

In 1990 Egypt signed and ratified the major human rights treaty relating to children, the Convention on the Rights of Child. As explained in Chapter Six, this means that Egypt, like the other ratifying countries, made a legal commitment to educate parents about the advantages of breastfeeding. Families should be well informed about the differences between breastfeeding and feeding with formula, especially the differences in their impacts on the health of both the infant and the mother.

Egypt's government should act decisively to protect the well-being of Egypt's infants. It would be unwise to end the formula subsidy program suddenly, but it could be carefully phased out. It should be replaced with good support for breastfeeding, including good information about formula and other foods for infants and young children. Appropriate food, vitamins, and micronutrients could be provided to lactating mothers. When the health and economic benefits of breastfeeding are well understood, these changes would be welcomed by Egypt's families.

CHAPTER FOUR

UNITED STATES

A note in my files reminds me that in February 1997 I phoned someone at the WIC head office in the United States Department of Agriculture to ask about the WIC program. My note says, "He does not believe formula displaces breastmilk. He believes WIC mothers choose formula, and WIC simply responds to their choice. He believes the WIC population is especially prone to choosing formula feeding." This view from WIC's head office, that WIC simply responded to mothers' choices, troubled me. My interest in the program has persisted since then.

The mission of the United States' Special Supplemental Nutrition Program for Women, Infants, and Children, commonly known as WIC, is "To safeguard the health of low-income women, infants, and children up to age 5 who are at nutrition risk by providing nutritious foods to supplement diets, information on healthy eating, and referrals to health care (Women, Infants and Children 2016)."

WIC helps many poor and low-income families. Low income families are defined as those under 185 percent of the poverty line. In addition, families eligible to receive benefits from certain other federal programs are categorically eligible for WIC services. After a two-year trial period, the program began with about 88,000 participants in 1974. It has grown steadily, reaching more than eight million in 2015 (United States Department of Agriculture 2016b).

The program serves more than half the infants in the United States, and about a quarter of all children less than five years of age. It serves about two million infants each year. The program provides food and health services for children up to the age of five, but formula is provided only to the child's first birthday. More than 90 percent of the infants in the program get some formula from WIC (Patlan and Mendelson, 2016, 32).

WIC distributes a large volume of infant formula at no cost to its participating families. In 2004-2006, 57-68 percent of all infant formula used in the U.S. was provided through the WIC program (Oliveira, Frazão, and Smallwood 2010, 1).

WIC's total program costs peaked at more than seven billion dollars in 2011and declined to $6.17 billion in 2015 (United States Department of Agriculture 2016b).

WIC is managed under detailed federal laws and regulations (United States Department of Agriculture 2016d, 2016g). It is a layered organization, based in the Food and Nutrition Service of the U.S. Department of Agriculture. State-level offices and similar offices serving Indian tribes and U.S.-affiliated territories administer WIC's operations. The state offices oversee local offices distributed through their cities and towns. Typically, the states' health departments manage the state-level operations. All program offices operate under a common framework of WIC rules, but they have some latitude to do things their own way. Some activities undertaken in one office might not be undertaken in others.

Apart from these governmental activities at various levels, nongovernmental organizations have been created to provide research, advocacy, and educational support for the WIC program. There is a nongovernmental National WIC Association, and several regional ones. They are funded privately, by grants from the food industry and other sources.

WIC'S BREASTFEEDING SUPPORT

WIC has a vigorous breastfeeding support program that operates in parallel with its formula distribution program. It has done some very good things, such as redesigning the food packages to promote and support breastfeeding. Formula amounts are tied to feeding practice and the age of the infant, complementary foods are delayed to 6 months, and fruit juice has been eliminated. Fully breastfeeding mothers receive a greater variety and a larger quantity of food. They also receive baby food meat because of concerns that the iron and zinc content of breastmilk might be inadequate for older infants. Bonuses are provided to WIC in states that show high increases in rates of exclusive breastfeeding. Some WIC offices have established breastfeeding peer counseling programs. Many employ International Board Certified Lactation Consultants to support new mothers. Many WIC offices offer excellent breastfeeding education programs.

Some WIC offices impose stringent limits, providing formula to participants only when medically indicated. Some provide formula to any participant who asks for it. Some local WIC agencies maintain Baby-Friendly policies and practices in their clinics. Several WIC policy statements refer to the International Code of Marketing of Breast-milk Substitutes (discussed in Chapter Five) for guidance, even though there is no federal law requiring adherence to it.

WIC's breastfeeding support has become more effective over time. However, it is useful to compare the magnitude of WIC's breastfeeding support effort and the magnitude of WIC's formula distribution. WIC spends more on formula than any other food, almost a billion dollars in 2010. In 2011, WIC spent $162,126,986 on breastfeeding support. About *six* times as much of WIC resources went into its formula program as went into its breastfeeding support program.

WIC pays only about eight percent of the wholesale value of the formula. Conservatively, the retail value of the formula to WIC participants is least ten times what WIC spent on the product. Thus, in crude monetary terms, from the perspective of WIC participants, the formula provided was worth at least *sixty* times as much as the breastfeeding support.

An earlier study reported similar findings:

> According to WIC, "A breastfeeding mother and her infant shall be placed in the highest priority level." Despite this statement and others that support breast-feeding, WIC allocates only 0.6% of its budget toward breastfeeding initiatives. Formula expenses accounted for 11.6% ($850 million) of WIC's 2009 expenses. The inconsistency between WIC's policies that encourage breastfeeding vs. practices that favor formula begs further examination. Research shows consistent success with peer counseling programs among WIC partici-pants; however, little money is budgeted for these pro-grams. Rebates included, WIC spends 25 times more on formula than on breastfeeding initiatives. (Baumgar-tel and Spatz 2013)

Comparing the size of WIC's formula distribution and breastfeeding support efforts is interesting, but it is really not important. We should not accept harms done to one group of WIC's participating families just because good is done for others. We would not allow a hospital to do harm to patients in one part of the building just because it provides good benefits to patients in another part.

PROCUREMENT POLICY AND PRICES

WIC obtains infant formula as follows:

> WIC uses a competitive bidding process under which infant formula manufacturers offer discounts, in the form of rebates, to state WIC programs in order to be selected as the sole formula provider to WIC participants in the state. WIC purchases of infant formula account for more than half of domestic infant formula sales.
>
> The competitive bidding process yields $1.5 billion to $2 billion a year in rebates, with WIC paying on average only 8 percent of the formula's wholesale cost. As a result of these savings, WIC's cost to the federal government is much lower than the full retail value of WIC benefits for program participants. (Center on Budget and Policy Priorities 2015)

Winning a WIC sole-source contract with WIC has a significant impact on infant formula manufacturers' market share in supermarkets:

> The impact of a switch in the manufacturer that holds the WIC contract was considerable. The market share of the manufacturer of the new WIC contract brand increased by an average 74 percentage points after winning the contract. Most of this increase was a direct effect of WIC recipients switching to the new WIC contract brand. However, manufacturers also realized a spillover effect from winning the WIC contract whereby sales of formula purchased outside of the program

also increased. (Oliveira, Frazão, and Smallwood 2010, 18; also see Davis and Olivera 2015; Ginty 2011; Oliveira and Frazão. 2015a, 35)

Many families don't get enough formula from WIC to meet their infants' needs. When they exhaust their free supplies for the month, they can purchase formula in the regular open market. If they stay with the brand they got free from WIC, they would spend more than they would by buying generic "store brand" formula.

WIC participants exchange their vouchers or use their EBT (Electronic Benefit Transfer) cards for free formula at retail stores. Remarkably, WIC then compensates the retailers based on the retail prices they charge to those consumers who pay with cash, not the wholesale prices the retailers pay their suppliers. WIC does not question the retailers pricing policies. This means the retailers have a strong incentive to bump up the prices they charge. This benefits the shop owners and hurts the paying customers, the ones who do not have WIC vouchers.

WIC's formula distribution leads to higher retail prices for the products:

> Over half of the infant formula sold in the United States is purchased with WIC benefits. By providing low-income families with free formula, WIC essentially replaces price-sensitive consumers from the infant formula market with price-insensitive consumers. As a result, both manufacturers and retailers raise their prices. (Oliveira and Frazão 2015b)

The retail price increases are higher for the name brand formulas that have WIC contracts. This increase affects WIC participants who are no longer eligible for free formula, former WIC participants, and all other purchasers of the branded formula that had the local WIC contract. As the retail prices of these branded formulas become extra high, the profit margins for sellers and manufacturers also become extra high.

The same pattern has been demonstrated for WIC foods other than formula:

> Participants receive the foods covered by WIC at no charge, and thus they lack the incentive to be price conscious in their purchase decisions. This creates the potential for WIC vendors to charge higher prices for WIC-eligible products, relative to comparable products that are not WIC eligible. (Saitone, Sexton, and Volpe 2015)

The government's unquestioning payment based on the retail prices to the storekeepers echoes a much broader pattern in the United States: "The history we see over and over again is that when the government steps in as a guaranteed payer without regard to price, it will be taken advantage of (Alonso-Zaldivar 2016)."

There are comparable rules for Medicare Part D, under which the government is forbidden by law to negotiate lower drug prices. This helps to explain the rapid inflation of drug prices in the U.S (Lee, Gluck, and Curfman 2016).

WIC's procurement rules are simple:

> Each WIC State agency, or group of agencies, awards a contract to the manufacturer offering the lowest net wholesale price, defined as the difference between the manufacturer's wholesale price and the State agency's rebate. (Oliveira, Frazão, and Smallwood. 2010)

This means WIC cannot evaluate alternative formula products, and it cannot negotiate. As Maureen Minchin explains:

> . . . legislation in 2004 removed WIC's ability to determine which formulas it wanted from a tendering company; companies could decide what formula they offered to WIC, at what price, provided the formula met the specifications for WIC use. If companies offered only the most expensive or novel brands, WIC had no choice but to become the inadvertent marketer of such products, without proof of the many advertising claims. In every state, WIC acceptance of any infant formula tender means massive increases in sales of the chosen brand. (Minchin 2016)

The similarities between the WIC procurement rules and those in Medicare are not surprising, given the infant formula industry's historic connections to the pharmaceutical industry (Samuels 1993; Wattana 2016).

These constraints in WIC's procurement practices benefit the sellers and lead to increasing costs to taxpayers, WIC families, and families outside the WIC program. The pattern raises questions about possible monopolistic practices, already raised in several lawsuits (Minchin 2016). Why do so few manufacturers enter what appears to be a highly profitable industry?

QUESTIONABLE ADDITIVES

Under this procurement system, if all the bidders raise their prices because some new additive has been introduced, WIC cannot challenge their claims about the supposed benefits from those additives (Neuberger 2010). Many additives are offered by manufacturers to make their brands more attractive, but there are serious questions about their effectiveness in relation to the claims made for them. Few additives have been tested properly by independent research agencies to determine whether the additives do in fact provide the claimed benefits, and whether the benefits are sufficient to warrant the added costs.

The Food and Drug Administration has taken slow, tentative steps toward insuring the manufacturers claims are valid (United States Food and Drug Administration 2016).

Many claims about the benefits of additives in infant formulas are questionable and some are misleading (Belamarich, Bochner, and Racine 2015; Kent 2014a; Lampl, Mummert, and Schoen 2016). Although it is never clearly stated, most claims for new versions of infant formula are based on comparisons with old versions. The manufacturers avoid making comparisons between formula and breastmilk, especially in terms of their functions.

In the United States overall (not just WIC), "Sales of DHA/ARA-supplemented formulas increased rapidly, and by 2004 they accounted for 69 percent of all sales. By 2008, DHA/ARA-supplemented formulas accounted for 98 percent of all formula sales (Oliveira, Frazão, and Smallwood 2010, 9)." Such additives play a distinctive role in shaping prices in the WIC context:

> All rebate contracts in effect in December 2008 were based on formulas supplemented with the fatty acids docosahexaenoic acid (DHA) and arachidonic acid (ARA), whereas most of the previous contracts were based on unsupplemented formulas. Because *wholesale* prices of DHA/ARA-supplemented formulas are higher than wholesale prices of unsupplemented formulas, wholesale prices of infant formula increased more in States that switched to the more expensive DHA/ARA-supplemented formula in their contracts that were in effect in December 2008. (Oliveira, Frazão, and Smallwood 2010, iii; also see Kent 2014a)

According to a 2010 report from the nongovernmental Center on Budget & Policy Priorities, "WIC appears to be spending more than $90 million extra annually — or more than 10 percent of its total spending on infant formula — to provide formulas with ingredients that neither USDA nor the FDA has assessed with regard to their benefits." It makes no sense to have "WIC spend extra taxpayer funds on ingredients without considering whether they provide health or development benefits. This position is not responsible (Neuberger 2010, 2)."

Additives to formulas can lead to systematic formula price inflation: "Thus, a troubling and costly cycle is beginning to affect WIC's bottom line, in which the 'designer' formulas that carry higher wholesale (and retail) prices become the standard, increasing WIC costs and requiring Congress to appropriate more funds for WIC to pay for them (California WIC Association 2010)."

When the manufacturers offer only formulas with questionable new additives when bidding for WIC contracts, the government's procurement rules in effect require that infants in the WIC program children must consume them. That formula that will be costly to the families when the free supply runs out.

PROMOTING FORMULA

Usually, when a family is accepted into WIC, there is no difficulty in getting formula. WIC provides some special formulas for cases that show special medical needs, but for ordinary formula there is no requirement to claim such needs. The basic policy is that for participants, "WIC State agencies provide infant formula for mothers who choose to use this feeding method (United States Department of Agriculture 2016d)." With this easy access, it is no surprise that, "Prenatal WIC participation is associated with a greater likelihood of providing babies infant formula rather than breastmilk after birth (Ziol-Guest and Hernandez 2010; also see Ryan and Zhou 2006)."

Experts on WIC acknowledge:

> In general, despite WIC's strong policy and operational emphasis on promoting breastfeeding, mothers participating in WIC have been less likely than non-participating mothers to breastfeed their infants. Whether this is because providing free infant formula creates an incentive for formula feeding or because mothers who are less likely to breastfeed are also more likely to participate in WIC remains unclear. (Carlson and Neuberger 2015, 11)

Either way, there is no doubt that the offer of free formula is an incentive to feed with formula. It may be that WIC enrollees tend to be women who would be less likely to breastfeed even in the absence of the WIC program, and might have different attitudes about formula feeding and breastfeeding. Nevertheless, making formula available for free certainly makes it more likely that mothers will use formula, regardless of what their initial inclinations might be.

The fact that many families drop out of the WIC program after their infants' first birthdays supports the idea that many come to WIC mainly because of the free formula. In 2012, WIC served 51 percent of United States infants under one year of age, but only 28.4 percent of children under five years of age (Oliveira and Frazão 2015). The proportion of children enrolled as WIC participants decreases steadily after their first year (Center on Budget and Policy Priorities 2015).

The California WIC Association plainly acknowledged:

> Duration of breastfeeding beyond the first few months
> is also rare in the WIC population: in California, only
> about 18% of WIC mothers are still breastfeeding af-
> ter the first three months. Using formula undermines
> breastfeeding because it interferes with a mother's abil-
> ity to establish her milk supply. (California WIC Asso-
> ciation 2010)

In providing free infant formula through WIC, the government
promotes formula use. This goes against WIC's own
recommendations and those of many other government agencies
(United States Department of Health and Human Services 2011;
United States Centers for Disease Control and Prevention 2016a,
2016b, 2016c) and also non-governmental agencies (American
Academy of Pediatrics 2012; National Alliance for Breastfeeding
Advocacy 2016; United States Breastfeeding Committee 2016).
More are listed by the Centers for Disease Control (United States
Centers for Disease Control 2016b).

The government's large-scale distribution of free infant formula
certainly muddles its breastfeeding support message. This is
reflected in systematic interviews with mothers:

> Participants view the WIC program in a contradictory
> manner. They see it as highly supportive of breastfeed-
> ing, but also as a promoter of infant formula. The ex-
> panded food package for mothers is not valued, but free
> supplemental formula is highly valued. Misinformation
> about breastfeeding pervades the healthcare system, and
> exclusive breastfeeding is not promoted as an important
> health goal. (Holmes et al. 2009)

THE ADDICTION MODEL

Some might believe that manufacturers provide formula at sharply reduced prices because so many families are needy. They have low incomes and their mothers cannot or will not breastfeed them, so the government does what needs to be done. But if the formula companies provided the formula out of compassion for the poor, they would seek tax deductions for these generous gifts. They do not do that. The WIC contracts are business transactions, not acts of charity.

Some people assume that United States taxpayers bear the high cost of this gift to the needy. But the infant formula is obtained through a system of competitive bidding under which manufacturers pay rebates to WIC that in 2013 averaged 92 percent of the wholesale price. WIC pays around a billion dollars a year for infant formula. That number is large because the total quantity purchased is large, not because the unit price is high. WIC pays very little for each can of formula.

The rebates have been used to reduce WIC food costs paid by the government. In Fiscal Year 2014, for example, federal saving from infant formula rebates amounted to $1.8 billion (Center on Budget and Policy Priorities 2015). The savings allowed WIC to provide its services to more families. In 2008 about a quarter of WIC's caseload was based on funding from the rebates (Oliveira, Frazão, and Smallwood 2010, 1).

The question then is, why are the manufacturers so generous? The large rebates are possible partly because the cost of manufacturing formula is low, so it is easy for manufacturers to offer large discounts. The powdered cow's milk used as the basic ingredient in most formula is an inexpensive bulk commodity. The fact that the product can be made for far less than the price at which it can be sold is important. However, the *long-term* benefits to the manufacturers are even more important.

In the United States the formula market is dominated by Abbott, maker of the Similac product line, Mead Johnson, maker of the Enfamil line, and since the late 1980s, Nestlé/Gerber, maker of the Good Start line. Outside WIC, the name brand companies provide generous samples of infant formula and other baby foods through maternity hospitals, obstetricians' offices, and other channels. The following image shows the types of gift pack women typically receive during pregnancy and at hospital discharge.

Figure 4-1. Free Samples for New Mothers.

Largely because of the Baby Friendly Hospital Initiative, there has been progress in limiting the distribution of such hospital discharge packs (Strader 2016; UNICEF 2016c), but the progress has been uneven.

Here is one mother's insight on why the manufacturers provide free samples:

> Parents who buy infant formula and stick with the brand the hospital distributed are paying tons of money for those "free" samples — $700+ per year over the cost of generic formula. The reason the formula companies include all these freebies in their advertising budgets is because it generates huge profits! (KellyMom 2015)

Others put it this way:

> *There is always free cheese in a mousetrap*. Free samples of infant formula, coupons, and hospital discharge packs are pervasive in the United States. (Lazarov and Evans 2000, 18)

The distribution of these free products promotes the use of infant formula. More importantly, it has the effect of endorsing specific brands. Parents are likely to assume that authoritative health professionals choose the brands based on the quality of the product.

The distribution of free formula through the WIC program serves the manufacturers' interests in much the same way as the hospital discharge packs. The provision of free formula through WIC is at a far larger scale, for each family, and also for the number of families affected. Both methods of distribution draw families into favoring particular brands of formula, and in the longer run, also favoring those same brands of follow-on baby food.

Considering only the short-term impact of the increased retail prices underestimates the benefits to the manufacturers of gaining WIC contracts. There is also a considerable long-term benefit because people who start with a major brand of infant formula are likely to stay with it when they buy follow-on formula and other baby foods in the following years. Though not as visible as the short-term benefits, these long-term benefits to the manufacturers are important in their calculations (Kent 2012b).

Figure 4-2. Viceroy Cigarettes Distributed by
Aloha Airlines in the 1950s
Source: https://www.etsy.com/listing/52245381/vin-
tage-1957-viceroy-cigarette-samples

Only manufacturers of the well-known brands bid for WIC contracts. Other manufacturers such as PBM Nutritionals (GoodGuide 2011) that produce generic or "store brands" do not bid. Since they do little advertising and do not have well-known lines of baby food that children might use over several years, the manufacturers of generic formula cannot build up a long-term loyal following by distributing free samples.

The analogy of cigarette samples explains the system. Older people may remember when little packs of four cigarettes were given away in airplanes and various public places. The object was to draw people not just to smoking, but to particular brands. It was worthwhile for the companies to hand out free samples in the hopes that people would become their customers for a long time.

Others have noted the similarity:

> [A] common thread between formula feeding and smoking is consumption patterns. They both follow the addiction model: It's easy to get hooked, and then you've gotta have it . . . During World War II, soldiers were issued with free cigarettes, courtesy of the tobacco companies, whereas today formula companies gatecrash maternity wards. (The Alpha Parent. 2015)

As Virginia Thorley put it, "soldiers who were previously nonsmokers began smoking since it cost them nothing, seduced into smoking in the same way as breastfeeding mothers are induced to use AIM [artificial infant milk] (Thorley 2015)."

Just as no formula maker is motivated to hand out samples of generic formula, no cigarette maker is motivated to hand out generic cigarettes. This helps to explain why so few formula manufacturers bid for WIC contracts.

The distribution of free infant formula and other products to new mothers, whether through maternity wards in hospitals, through frequent mailings to new mothers, through doctors' offices, or through the WIC program is driven by commercial interests like those that once drove the distribution of free cigarettes.

Most addictions are linked to the chemistry of substances such as recreational drugs, tobacco, alcohol, sugar, salt, and fat (Moss 2013). The attachment to infant formula or to brand names is different, but it can work in a comparable ways and lead to comparably harmful results. Infant formula is special because it is many people's first ultra-processed food, a point of entry or gateway to other ultra-processed foods later in life (Kent 2012b; Monteiro et al. 2016).

The long-term effect might be inter-generational, as suggested by Maureen Minchin's observation:

> By as young as 24 months infants predominantly fed formula for the first 6 months have two and a half times the risk of being obese, as their usually bottle-fed parents have been before them. (Minchin 2016)

Children do not become hooked on specific brands, but their parents do. One leader in the business observed, "People don't switch brands in baby food unless their baby is not well. Brand loyalty is passed on from mother to daughter; price is never an issue (The Economist 2006)."

Attracting women to a brand can benefit manufacturers well beyond the time when the supply of free formula from WIC comes to an end at the child's first birthday. For some, that supply shortage can occur even sooner. As one observer put it, "the habits and brand loyalty formed by the WIC rebate system can hook women on paying retail prices for formula once their stipend runs out each month (Ginty 2011)."

The brand loyalty is likely to persist and draw parents to follow-on formulas and other baby foods marketed under the same brand name.

The manufacturers that bid for WIC contracts do that not because the contracts are highly profitable in themselves or because of their deep compassion for the poor, but because the WIC program attracts people to their brands.

The cigarette companies had to pay people to distribute their free samples, but in the United States the infant formula manufacturers get the national government to do the job for them. The companies don't have to pay any of the labor costs relating to the distribution of their products.

The WIC contracts are so lucrative, the companies might find it in their interest to give rebates of 100 percent or, to simplify matters, just give WIC the products free. Having the formula companies provide the product to WIC at no cost would eliminate the task of managing the rebate program.

Business people know that handing out products for free can be a good way to recruit regular customers. Having it handed out by prestigious professionals such as health care workers or by agencies of the government is even better. The major formula companies have had a great deal of experience with that practice, handing out their products in maternity wards and obstetricians' offices and through the mail. Now they do it through their arrangements with WIC, with the government carrying the costs.

COMMODITY SUBSIDIES

The focus here has been on the United States government's provision of free infant formula in huge quantities. In reflecting on this, we should recall the WIC program's relationship to the government's long history of subsidizing farmers who produce major commodities such as corn, soybeans, wheat, rice, dairy, and livestock. A physician at the Centers for Disease control observed:

> In the U.S. and many other places, an excess of subsidies in these areas ends up leading to a conversion into

foods like refined grains and high calorie juices, soft drinks with corn sweeteners and high fat meats. (Oaklander 2016)

A group of physicians made similar observations, and pointed out that one of the means of subsidizing commodity producers is through federal purchase programs:

> The federal government, through USDA, provides additional support to livestock and crop producers by purchasing agricultural products for use in the National School Lunch Program, the Special Supplemental Nutrition Program for Women, Infants and Children (WIC), and the Emergency Food Assistance Program (TEFAP).

That group also explained:

> The USDA refers to fresh fruits and vegetables as "specialty crops." Specialty crops do not receive subsidies. In fact, farmers who participate in commodity subsidy programs are generally prohibited from growing fruits and vegetables on the so-called "base acres" of land for which they receive subsidies. (Physicians Committee for Responsible Medicine 2016; also see Union of Concerned Scientists 2016)

In light of the USDA's historic interest in finding uses for overproduced commodities, it is not difficult to understand why the WIC program is lodged in the United States Department of Agriculture and not the Department of Health. WIC's roots trace back to the USDA's Commodity Supplemental Food Program, established in 1969. It provides commodities to feed low income pregnant women, infants, and children up to age 6 (Institute of Medicine 1996, 30; Oliveira et al. 2002, 7).

This misplacement of a program whose primary mission is health-centered leads to peculiar phenomena. It explains why the Department of Agriculture, rather than the Department of Health, took on the task of assessing the health effects of soy-based infant formula (Badger 2009; United States Department of Agriculture. 2004)." It is not clear how a USDA study could assess a long-term health effect such as sexual dysfunction by examining children only up to the age of six. Agricultural agencies are interested in promoting the use of soy. They don't have much expertise in assessing the health impacts of infant foods (Kent 2011, 25-27). Why should agriculturalists have any role in assessing the health impacts of infant formula?

WIC'S OPTIONS

WIC'S large-scale distribution of infant formula at no cost to families serves large agriculture interests and the interests of manufacturers and sellers of infant formula, but at the same time it harms families. This needs to be studied so the situation can be properly diagnosed and remedied. There are things WIC could do to increase its effectiveness in fulfilling its mission "to safeguard the health of low-income women, infants, and children up to age 5."

Research

A good place to start would be to seek fairer and more comprehensive research studies on the impacts of WIC's activities. Both parents and governments need to make well informed policy choices based on a clear understanding of the likely consequences of those choices.

In 2004 the United States government published a report on how the marketing of infant formula might discourage breastfeeding. It never raised the question of whether WIC's formula program itself might discourage breastfeeding. The study's only recommendation was to call on the Secretary of Agriculture to "educate all states about its policy restricting the use of the WIC acronym and Logo . . . (United States Government Accountability Office 2006, 35)."

WIC has had research done on its breastfeeding program (Mathematica Policy Research 2015; United States Department of Agriculture 2016a), but nothing of similar depth on its formula distribution program. There are no systematic studies that compare the two programs. There is no research that compares the health impacts of the two different approaches to feeding infants, despite the centrality of health in the agency's core mission.

Similarly, no systematic research has been done to compare the impacts of the two approaches on the budgets of families in the WIC program, or those who have left the program, or those who have never been in the program.

The major international agencies concerned with infant feeding agree that optimal breastfeeding practices require initiation within the first hour after birth, six months of exclusive breastfeeding, and continued breastfeeding for up to two years and beyond, together with other foods. One indicator commonly used to assess progress in breastfeeding practices is the proportion of infants who are exclusively breastfed for at least five months. The report from Mathematica Policy Research had no clear figures to report. Instead it discussed the lack of a sound procedure for estimating the rate of exclusive breastfeeding among WIC participants.

WIC has been providing breastfeeding data annually since 2011, accessible through their website on Breastfeeding Promotion and Support (United States Department of Agriculture. 2016f). WIC collects data on breastfeeding "outcomes" such as initiation, duration, exclusivity, and intensity. The significance of these reports is not always clear. For example, the report for Fiscal Year 2015 said:

> The data indicate an increase of 0.5 percent in the number of total WIC infants reported as Breastfed from 30.4% in FY 2014 to 30.9% in FY 2015. This increase in breastfeeding rates represents the successful efforts of WIC State and local agencies in continuing to provide quality breastfeeding services, including peer counseling, to WIC participants. (United States Department of Agriculture 2016a, 3)

An increase of half a percent hardly seems something to boast about. Also, the presentation is not clear on how "Breastfed" is defined. Apparently, it refers to whether the infant has ever been breastfed, so even a single instance of breastfeeding would be counted as a success. It is far more important to know what proportion of WIC's infants met the goal of having been exclusively breastfed from birth and continued for five or, some say, six months (Drago 2009, Figure 1).

Only 12.9 percent were "fully" breastfed, up 0.1 percent over the previous year. The meaning of "fully" was not explained, and the indicator for it was not described in this document.

WIC offers different food packages, distinguishing between fully breastfeeding, partially breastfeeding, and fully formula feeding. "Fully breastfeeding food packages are for mothers and their babies who do not receive formula from WIC and are assumed to be breastfeeding exclusively (United States Department of Agriculture 2016e)." Categorizing women as exclusively breastfeeding based on which WIC food package they choose is misleading. Some women might choose because they prefer the food package that goes with that choice. This approach is not likely to yield data comparable with the data on exclusive breastfeeding reported by agencies such as the United Nations Children's Fund and the World Health Organization.

Data on initiation, duration, exclusivity, and intensity of breastfeeding can be interesting, but data on health indicators would be of greater value. If health indicators were used, it would be possible to compare health outcomes of breastfed infants with health outcomes of formula fed infants, controlling for possible confounding factors. Unfortunately, WIC does not publish data that would support comparisons of health outcomes with different methods of feeding.

Well-designed policy-oriented research might show that shifting WIC funds from purchasing formula to supporting breastfeeding more vigorously would yield significant health benefits and at the same time make good economic sense for affected families. To support this research, it would be helpful to have details of infant feeding practices from birth in all health records.

It is difficult to generalize about the impacts of food subsidy programs because they are diverse and there are many impediments to doing research on that topic (Black et al. 2012). However, the WIC program has excellent opportunities to facilitate good research on its programs.

Several studies have shown that the WIC program overall (not just its formula distribution program) has been effective in reducing child mortality, improving children's health, and reducing their health care costs, especially their Medicaid costs (Black et al. 2004; Carlson and Neuberger 2015; United States Department of Agriculture 2016h). Some studies are of questionable value (Besharov and Germanis 2000).

Why is it that in a study of obesity in 2- to 4-year-old children in the WIC program, no distinction is made between those who were predominantly formula fed and those who were predominantly breastfed (Pan et al. 2016)"? It would easy enough to do health impact studies with clear distinctions made between those two groups. This seems to have been systematically avoided in studies on WIC.

If WIC provided better information on the feeding practices of its participants, independent researchers could make comparisons between the two groups, carefully qualified by an appreciation of the possible confounding factors. Certainly, a program whose core mission is to improve health needs to know how each component contributes to fulfillment of the mission.

More Comprehensive Cost-Benefit Analyses

The results of cost-benefit analyses depend on how they are framed. Free or highly discounted infant formula might look like a good bargain to new parents. But if broader considerations are considered, such as health care costs or the costs of long-term commitment to a famous brand, the findings are likely to be different. Good food may be costly, but it can be viewed as a sound investment.

Researchers should give attention to the health and economic costs that are presently externalized and thus disregarded. Some assessments of the WIC program focus narrowly on direct costs to government, and fail to estimate or even recognize other important impacts. For example, the benefit WIC contracts bring to manufacturers in building long-term brand loyalty is rarely mentioned. Economic arrangements that seems cost effective for government are not necessarily cost effective for WIC's participants (Carlson, Greenstein, and Neuberger. 2015).

Just as American soldiers were glad to get Coca-Cola for only five cents during World War II, the United States government itself must have been happy to get these cheap drinks to pass on to the soldiers. It looked like a win-win situation. The reports about that program present the distribution of over five billion bottles of Coca-Cola as a hugely generous and patriotic act of the company (Digger History 2016; Journey Staff 2012; Mooney 2008).

The reports on the Coca-Cola distribution program make it look like everybody won because they used a narrow frame of analysis, ignoring long-term impacts. The reports don't discuss how Coca-Cola's generosity turned out to be a brilliant marketing maneuver, helping to establish the drink in American culture and in many other parts of the world, impacts that have lasted well beyond the war's end. Many people are now having second thoughts about those long-term impacts. Like new mothers, governments can be drawn in by free or highly subsidized goods, even when those goods might not be the best choice.

More should be done to guide decision-making not only by high-level policymakers but also by WIC participants themselves. WIC could do more to support parents in making well-informed choices about how they will feed their infants by providing them with appropriate user-friendly information. They need to know what impacts to expect with different methods of feeding.

Just as it would be useful for WIC to study how health and other impacts might be related to infant feeding methods, such studies would be of interest to many other agencies as well. If hospitals and health insurance agencies kept track of how infants were fed, they would be able to explore how those data relate to other relevant data, and develop better policies. Health maintenance organizations that treat people through their life spans would be in a particularly good position to assess long-term impacts.

Phase Out Formula Program

Beyond research, bolder lines of action should be considered.

The National WIC Association listed barriers that "interfere with WIC's breastfeeding promotion efforts and contribute to reasons why a mother may introduce formula (National WIC Association 2014)". They listed several important concerns, but somehow failed to mention that WIC's offer of free formula might be a factor. Perhaps they had not read the California WIC Association's observation that "Infant Formula Competes with Breastmilk":

> Infant formula companies battle for market share against a unique product: breast milk, a living food that contains hundreds of active biological substances that cannot be manufactured and are not present in infant formula. As breastfeeding rates have slowly and steadily increased, particularly among low-income women, the formula industry has grown more aggressive in its attempt to regain market share, particularly by pushing formula supplementation (i.e., combining breastfeeding and formula feeding. (California WIC Association 2010)

WIC could phase out its infant formula distribution program over a period of years (Kent 2016; Tuttle 200; Wood 2011). It could be replaced with cash or food grants of comparable value. Appropriate food, vitamins, and micronutrients could be provided to lactating mothers. WIC is intended to be a supplemental program, not a source for the entire diet of women or children. Families would still be free to use their own funds to purchase infant formula. People with low incomes could be helped in many ways without promoting the use of infant formula.

During the phase-out period, WIC could require its participants to pay for infant formula, but at reduced prices. When there is a cost for formula, it would not be quite as easy a choice as when it was completely free (Greiner 2011).

Britain's Healthy Start program provides vouchers that can be exchanged for various foods, including infant formula. WIC could follow that practice, offering vouchers of specific cash value that could be used for many different types of groceries. With that approach, there would be no distinct incentive to use infant formula. If WIC families had good alternatives regarding how they allocate their subsidy, many would skip the formula and choose something else instead. It has been shown that WIC participants will move to healthier food options when given a choice (Ettinger 2016; Strom 2016).

Only Generic Formula

If WIC does provide infant formula, it could make contracts only for generic ("store brand") formula that meets specific nutritional and safety requirements. Brand information would not be conveyed, and the rebate program would be ended. The net cost to WIC might be higher, but the net cost to families that do not get free formula would be lower (Oliveira, Frazão, and Smallwood 2010; Oliveira and Frazão 2015a.). If generic formula was used, rather than the well-known brands, WIC contracts would not bump up market prices as much, and families would be more likely to choose cheaper generic formula after their supply from WIC came to an end.

Offer Human Milk

The U.S. government could reposition WIC by having it promote breastfeeding more vigorously. When that does not work, WIC could offer the alternative of banked or shared human milk. Human milk could be costlier to WIC than formula, but it would be better for children's health in both the short and long term. It would reduce their health care costs.

WIC's providing subsidized generic formula or human milk for WIC participants would increase the costs to the government, when compared with the current practice of using famous brands of formula. The makers of generic formula and the human milk banks are not going to offer rebates or discounts comparable to those offered by the big-brand manufacturers. If WIC's primary mission is to promote health, it should be willing to pay the costs for doing that.

The total amount spent by WIC for human milk could be limited by providing it for a smaller number of infants than are now provided formula. WIC's support for the development of a safe and effective system for providing human milk for WIC participants could drive down the cost of human milk for all infants.

The human milk option is discussed more fully in Chapter Seven.

Research strategies are discussed more fully in Chapter Eight.

CHAPTER FIVE

THE MARKETING CODE

The International Code of Marketing of Breast-milk Substitutes was adopted by the World Health Assembly in 1981 because of widespread concern that formula and other baby foods were being promoted in irresponsible ways (World Health Organization 1981). The problems were clear:

> Before the adoption of the Code, egregious marketing and promotion practices were rampant. Companies sponsored "pretty baby" shows and hired "mothercraft nurses" to visit homes and maternity wards. Radio jingles and print ads led to widespread consumer recognition of the products. The medical profession was targeted as a promotional ally; free samples at the health clinic and supplies from the maternity ward meant, to mothers, that the product was medically endorsed. Doctors and health facilities received various material benefits, everything from pens and key chains to cash payments and trips abroad, for their implied or explicit endorsement. (Margulies 1997)

The United Nations Children's Fund summarized the Code as follows:

> The World Health Assembly adopted the International Code of Marketing of Breast-milk Substitutes in 1981 to protect and promote breastfeeding, through the provision of adequate information on appropriate

infant feeding and the regulation of the marketing of breastmilk substitutes, bottles and teats. In subsequent years additional resolutions have further defined and strengthened the Code.

The code stipulates that there should be absolutely no promotion of breastmilk substitutes, bottles and teats to the general public; that neither health facilities nor health professionals should have a role in promoting breastmilk substitutes; and that free samples should not be provided to pregnant women, new mothers or families. All governments should adopt the Code into national legislation. Since 1981, 84 countries have enacted legislation implementing all or many of the provisions of the Code and subsequent relevant World Health Assembly resolutions. (UNICEF 2016d)

The International Baby Food Action Network (IBFAN) is a global nongovernmental organization devoted to the protection and improvement of infant feeding throughout the world. IBFAN and its International Code Documentation Centre do very effective work in monitoring the implementation of the Code worldwide (IBFAN 2016; ICDC 2016). The positions of countries on the adoption of the Code in their national law are described in periodic status reports, initially by IBFAN-ICDC alone, and now jointly with key United Nations agencies (World Health Organization, UNICEF, and IBFAN. 2016).

The Code and the subsequent resolutions by the World Health Assembly elaborating it are recommendations, not binding international law. In contrast to human rights treaties and other international instruments that become binding on nations through a signature and ratification process, the Code and the subsequent relevant resolutions are not open to signature and ratification. As shown in the status reports, many national governments have accepted the Code and adopted it in their national law. This domestic procedure can make these obligations binding within nations under their national law.

The Code focuses on the marketing of infant formula and other breast-milk substitutes by manufacturers and distributors. While the drafters of the Code knew that some national governments would not implement the Code effectively, they did not anticipate that governments themselves might become large-scale promoters of infant formula by providing it to families for free or for sharply reduced prices.

The Code was drafted based on the view that government and corporations are separate, with the government overseeing and regulating the corporations, but this model does not always apply. As shown in the preceding chapters, some national governments provide large quantities of free or subsidized formula to their people. In some places, such as China and Vietnam, the government is heavily involved in the formula business through direct ownership or investments in the manufacturing companies. The Code should be seen as applying to governments as well as to private businesses, especially where there is no sharp distinction between the two.

DEFINITIONS

Marketing

In the Code, marketing means "product promotion, distribution, selling, advertising, product public relations, and information services (World Health Organization 1981, 9)." These practices are troubling because infants and young children are likely to have worse health outcomes if they are fed with formula rather than being breastfed. This concern has motivated widespread support for the Code.

Infant Formula

The Code defines "infant formula" as a breast-milk substitute formulated industrially in accordance with applicable Codex Alimentarius standards, to satisfy the normal nutritional requirements of infants up to between four and six months of age, and adapted to their physiological characteristics." This definition is problematic because the meaning of "satisfy the normal nutritional requirements" is not clear. As explained in Chapter Eight, in the basic Codex Standards for Infant Formula, the ingredients are those "which have been proved to be suitable for infant feeding." How they could be "proved to be suitable" is not specified.

Breast-milk Substitutes

Infant formula is just one of several types of breast-milk substitutes. Distinctions have been made elsewhere (not in the Code itself) as follows: "standard infant formula (IF) for children less than 6 months, follow-on formula (FF) for 6 to 11 months old, and growing-up/toddler milks (TM) for children 12 months and older (Piwoz and Huffman 2015)." The World Health Organization regards all these categories as breast-milk substitutes, thus falling within the scope of the Code (World Health Organization 2013c)

The Code defines "breast-milk substitute" as "any food being marketed or otherwise presented as a partial or total replacement for breast milk, whether or not suitable for that purpose."

This suggests that a product is or is not a breast-milk substitute depending on the manufacturer's intentions, not on whether it does in fact tend to displace breast-milk. Manufacturers sometimes make their views on this very plain, as in this assertion in a formula advertisement:

> This product is not a breast milk substitute, but a grow-
> ing up milk specially suited to healthy young children
> from 1 years of age. (Milk Powder Solutions 2016)

In effect, this statement insists that the Code is not applicable to the promotion of this product.

It corresponds with the view presented in the Codex Standard for Follow-Up Formula. Its section 2.2 says:

> Follow-up formula is a food prepared from the milk
> of cows or other animals and/or other constituents of
> animal and/or plant origin, which have been proved to
> be suitable for infants from the 6th month on and for
> young children.

The document does not explain how this food could be "proved to be suitable".

Its Section 9.6 says, "The products covered by this standard are not breast-milk substitutes and shall not be presented as such (Codex Alimentarius Commission 1987)."

The United Nations Children's Fund took a very different position:

> It is UNICEF's view that a follow-on formula is as
> much a breastmilk substitute as infant formula. Indeed,
> follow-on formulas did not exist when the International
> Code of Marketing of Breastmilk Substitutes was ad-
> opted in 1981, and were developed by the baby food
> industry to try and get around the prohibition on pro-
> motion in the Code. This led the World Health Assem-
> bly to adopt Resolution 39.28 in 1986 stating that "the
> practice being introduced in some countries of provid-
> ing infants with specially formulated milks (so-called
> follow-up milks) is not necessary". Since, according
> to WHA [World Health Assembly], babies should be

> breast-fed for two years or beyond, any milk product marketed for use before that age will replace breastmilk and must be considered a breastmilk substitute for the purposes of the Code. (Alipui 2011)

In 2016 the World Health Assembly recommended:

> Products that function as breast-milk substitutes should not be promoted. A breast-milk-substitute should be understood to include any milks (or products that could be used to replace milk, such as fortified soy milk), in either liquid or powdered form, that are specifically marketed for feeding infants and young children up to the age of 3 years (including follow-up formula and growing up milks). It should be clear that the implementation of the International Code of Marketing of Breast-milk Substitutes and subsequent relevant Health Assembly resolutions covers all these products. (World Health Assembly 2016b, para 11)

Clearly, there is work to be done on harmonizing these views.

Samples and Supplies

The Code is particularly concerned about the distribution of free formula and other breast-milk substitutes. It defines "samples" as "single or small quantities of a product provided without cost."

"Supplies" means "quantities of a product provided for use over an extended period, free or at low price, for social purposes, including those provided to families in need."

Most national governments do not distribute free samples in single or small quantities, but some, such as Chile, Egypt, and the United States, distribute free formula as supplies.

SUBSEQUENT RELEVANT RESOLUTIONS

The Code is treated as a living document, clarified and expanded through a series of subsequent relevant resolutions from the World Health Assembly. These resolutions on the Code and other resolutions relating to infant and young child feeding are available online (World Health Organization 2008, 2016).

Several of those resolutions are particularly relevant to this book.

World Health Assembly resolution WHA39.28, adopted in 1986, says in section 2(6) that breast-milk substitutes that may be needed for children in hospitals should be "made available through the normal procurement channels and not through free or subsidized supplies."

Section 3(2a) says, "any food or drink given before complementary feeding is nutritionally required may interfere with the initiation or maintenance of breast-feeding and therefore should neither be promoted nor encouraged for use by, infants during this period (World Health Assembly 1986)."

World Health Assembly resolution WHA47.5, adopted in 1994, called on all states "to ensure that there are no donations of free or subsidized supplies of breast-milk substitutes and other products covered by the International Code of Marketing of Breast-milk Substitutes in any part of the health care system (World Health Assembly 1994; World Health Organization, UNICEF, and IBFAN 2016, 25). The status reports have not systematically monitored country performance on this issue. However, this book shows that several countries subsidize infant formula, violating the spirit of resolution WHA47.5.

In 2016, the World Health Assembly adopted resolution WHA 69.9, *Ending the Inappropriate Promotion of Foods for Infants and Young Children* (World Health Assembly 2016a). It addressed promotion of a wide range of baby foods, not just formula. It should be read together with the Codex Alimentarius Commission's Codex Standards for Follow-up Formula (Codex Alimentarius Commission 1987). The Codex recommendations are discussed further in Chapter Eight.

UNITED STATES POSITION ON THE CODE

Several parts of the Code are particularly relevant to the role of governments as promoters of formula:

Article 5.2 says, "Manufacturers and distributors should not provide, directly or indirectly, to pregnant women, mothers or members of their families, samples of products within the scope of this Code."

Article 6.2 says, "No facility of a health care system should be used for the purpose of promoting infant formula or other products within the scope of this Code.

Article 7.3 says, "No financial or material inducements to promote products within the scope of this Code should be offered by manufacturers or distributors to health workers or members of their families, nor should these be accepted by health workers or members of their families."

It is now clear that governments can violate the spirit, the principles, of the Code. Here we examine how it relates to United States practices. This type of analysis could be adapted and used in studies of practices in other countries.

The United States was the only country to vote against the adoption of the Code by the World Health Assembly in 1981. The story behind that refusal has been told by Morton Mintz (Mintz 2012).

The United States has taken no significant action to implement the Code. Some have a different view. According to the National Alliance for Breastfeeding Advocacy:

> The United States entered into a consensus agreement in 1994 endorsing the Code and other WHA resolutions up to that date. The US also joined all other nations in consensus agreement in 1996 and 2001 on support for the Code and the resolutions being considered at each time. (National Alliance for Breastfeeding Advocacy 2015)

However, United States support for subsequent World Health Assembly resolutions should not be interpreted as signifying its support for the Code itself. A comparison can be made with the fact that the United States has ratified two Optional Protocols linked to the Convention on the Rights of the child (OHCHR), although it has not ratified the Convention itself (OHCHR 2016). The United States support for these Optional Protocols associated with the Convention does not mean that it has made a legal commitment to uphold the Convention.

Regardless of the formal legal status of the United States position on the Code, there is no indication of the government's having made a serious commitment to it. The Code has not been made a part of national law and the government has not implemented it.

The National WIC Association has taken steps to align its practices with the Code (National WIC Association 2016), but as a nongovernmental organization, that has no legal impact on WIC itself or the United States government as a whole. Nevertheless, it is difficult to understand how the National WIC Association can support the Code and at the same time support WIC's formula distribution program.

Even if it has not made a commitment to the Code, it is meaningful to ask whether the United States government has violated it. The Code and the subsequent related resolutions describe a widely accepted set of principles, articulating a moral consensus accepted in much of the world. As one expert explained, though World Health Assembly recommendations are not binding, they "carry moral or political weight, as they constitute the judgment on a health issue of the collective membership of the [World Health] Organization (Shubber 1985, 30; also see Shubber 1998, 2011.)."

WIC's formula distribution program provides *supplies*, in contrast to the *samples* in gift packs often given by formula companies to new mothers. However, the two serve the manufacturers' interests in the same way, drawing the recipients into using formula and, beyond that, encouraging them to favor one brand.

Article 6.2 of the Code says, "No facility of a health care system should be used for the purpose of promoting infant formula or other products within the scope of this Code." WIC's core mission is to safeguard health, so it is part of the United States health care system. While WIC does not say its mission is to promote the use of infant formula, it would be difficult to argue that handing out large quantities of free formula does not have the effect of promoting its use. It is not only free, but it also appears to be endorsed by the government.

Article 7.3 says, "No financial or material inducements to promote products within the scope of this Code should be offered by manufacturers or distributors to health workers or members of their families, nor should these be accepted by health workers or members of their families." The rebates from the manufacturers to the WIC program constitute a strong financial inducement to the WIC program to promote the use of formula.

Some might argue there is no violation because (a) the United States government has never made a commitment to implementing the Code, (b) the Code is designed to limit the marketing efforts of corporations, not governments, (c) the large-scale distribution of free formula through the WIC program does not constitute the distribution of a few free samples of the sort envisioned in the Code. Reservations of this sort might be made by a lawyer in a court case, but the argument here is moral, not legal. The Code's guidance, affirmed through the widespread adherence to it around the world, tells us that it is wrong to distribute free formula in a way that induces people to use formula and other breastmilk substitutes.

Through WIC's formula distribution program, the United States government is a large-scale violator of the principles and the details of the International Code of Marketing of Breast-milk Substitutes.

The WIC program as a whole, and its breastfeeding support program in particular, do a lot of good for a lot of people. However, this does not diminish the need for critical analysis of WIC's infant formula program. The analysis here shows that the program tends to favor infant formula manufacturers and sellers at the expense of families in terms of their health and their household budgets.

Governments should subsidize children's health, not corporations' wealth. There is no sensible way to argue that a large proportion of the infants in any country should get free formula from their governments. National programs for distributing free infant formula should be phased out.

CHAPTER SIX

INFANT FEEDING AND HUMAN RIGHTS

Years ago, an article on the Code began by saying it is "part of the body of human rights instruments that have been developed under the auspices of the United Nations since the Universal Declaration of Human Rights was proclaimed in 1948 (Margulies 1997). This might have been wishful thinking. The Office of the High Commissioner for Human Rights and most human rights experts do not view the Code as a human rights instrument. However, with some work, an updated version could become a part of the global human rights discourse.

The human rights perspective is particularly important here because this book focuses on the actions of governments. Human rights are primarily about the rights of the people and the corresponding obligations of their governments to ensure the realization of those rights. Little of the ongoing human rights discourse has given attention to infants, despite their special needs and their great vulnerability. This chapter draws from my past writing on the topic (Kent 1997, 1998, 2001, 2004, 2005a, 2005b, 2006a).

LEGAL CONTEXT

The parties to infant feeding are, most obviously, the mother and the child. But there are others with interests and influence in the situation. There is the father, and there are siblings and the extended family. There are friends. There is the local community. There are doctors, nurses, and other health professionals. Employers are affected. The local government may be concerned in some way, and possibly the national government, and even some international organizations. And there are commercial interests.

Each of these parties has some interest in the infant feeding relationship. All of them may feel or claim that they have a common interest in the health and well-being of the infant, but they have other interests as well. The mother is and should be concerned with her own health and comfort. Siblings may be jealous because of the attention paid to the newcomer. Both father and mother may be concerned about the mother's being drawn away from work in the field or the factory, or from the work of caring for other family members. Older female relatives may want to influence the feeding process. Employers may be concerned with the ways in which breastfeeding takes the mother away from work, whether for minutes, hours, days, or months. They may be concerned that publicly visible breastfeeding will distract other workers.

Health care workers may be concerned with the well-being of the infant and the mother, but they also have other concerns. They may have only limited time and other resources for preparing the new mother for breastfeeding. Their incomes may be affected by the new mother's choice as to whether to breastfeed or not. Commercial interests may want to sell products, either to support breastfeeding, such as breast pumps or special clothing, or for alternatives to direct, breastfeeding, such as formula or special equipment for delivering expressed milk. Government officials may be swayed in different directions, depending on which of these parties has the greatest influence on them.

The phrase "breastfeeding as a human right" is ambiguous. It could refer to the rights of the infant or of the mother. Or we could think of them as bonded so closely that they are one, with no imaginable conflict between them. Perhaps that is usually the case, but there can be differences between them. They do not always "agree" on when to start or stop feeding. The infant may be insensitive to the inconvenience or even pain he or she may sometimes cause. The mother may also be unhappy about being drawn away from work, or from her husband, or from other children, or from rest. There can be real differences in interests between mother and child.

These parties can influence one another's decisions in many ways, through education, persuasion, money, affection. The infant may not appear to be influential, but its birth and its behavior affect the mother's hormones, and provide a positive stimulus for breastfeeding. The hormones of pregnancy also cause proliferation of the ducts and alveoli of the mother's breasts, in preparation for production of colostrum and mature milk. The delivery of the placenta after the birth of the infant leads to a drop in progesterone. That causes production of breastmilk within three to six days of the birth. Lactation is the natural and direct result of pregnancy and delivery.

Beyond that, the interests of the infant may have an impact if he or she is represented by surrogates, others who have some capacity in the situation and who speak and act on the infant's behalf. The infant has little direct power in the relationship. With this extreme asymmetry in the power relationships, it is especially important to articulate the rights of the infant.

THE HUMAN RIGHT TO ADEQUATE FOOD

As human beings, infants enjoy all human rights, to the extent that their limited capacities allow. The human rights of infants regarding their nutrition must be located within the broader context of the human right to adequate food in modern international human rights law and principles. The foundation lies in the Universal Declaration of Human Rights, which asserts, in article 25(1), that "everyone has the right to a standard of living adequate for the health and well-being of himself and his family, including food . . ."

All global human rights agreements discussed here are available through the website of the Office of the High Commissioner for Human Rights. They can be accessed on the Internet by going to www.ohchr.org and clicking on Issues/Human Rights Instruments.

The right to food was reaffirmed in two major binding international agreements. In the International Covenant on Economic, Social and Cultural Rights, which came into force in 1976, article 11 says, "The States Parties to the present Covenant recognize the right of everyone to an adequate standard of living for himself and his family, including adequate food, clothing, and housing . . ." It also recognizes "the fundamental right of everyone to be free from hunger . . ."

In 1999 the United Nations released its General Comment 12 on the right to food (United Nations Economic and Social Council 1999). Its explanation of the meaning of the human right to adequate food constitutes an authoritative contribution to international jurisprudence.

International human rights law has little to say about infant feeding. However, article 24 of the Convention on the Rights of the Child says, "States Parties recognize the right of the child to the enjoyment of the highest attainable standard of health . . . (paragraph 1)" and shall take appropriate measures "to combat disease and malnutrition . . . through the provision of adequate nutritious foods, clean drinking water, and health care (paragraph 2c)."

Paragraph (e) mentions breastfeeding. It says that States Parties shall take appropriate measures . . .

> To ensure that all segments of society, in particular parents and children, are informed, have access to education and are supported in the use of basic knowledge of child health and nutrition, the advantages of breastfeeding, hygiene and environmental sanitation and the prevention of accidents . . . (OHCHR 1990)

All countries except the United States have signed and ratified that Convention. This means they have made legal commitments to educate parents about the advantages of breastfeeding. Governments should ensure that families are well informed about the differences between breastfeeding and feeding with formula, especially their impacts on the health of both the infant and the mother. They should do this because it is a legal requirement, and because it is the right thing to do.

Apart from the Code and the subsequent related resolutions from the World Health Assembly, several non-binding international declarations, resolutions, and guidelines have helped to shape the emerging international consensus on the meaning of the human right to adequate food as it applies to infants:

- On August 1, 1990, the *Innocenti Declaration on the Protection, Promotion and Support of Breastfeeding* was adopted by participants at a meeting on Breastfeeding in the 1990s held at the Innocenti International Child Development

Centre in Florence, Italy. The declaration stated a variety of specific global goals, including the goal that "all women should be enabled to practice exclusive breastfeeding and all infants should be fed exclusively on breast-milk from birth to 4-6 months of age. Thereafter children should continue to be breastfed, while receiving appropriate and adequate complementary foods, for up to two years of age or beyond (Innocenti 1990)". In 1991 the UNICEF Executive Board passed a resolution (1991/22) saying that the Innocenti Declaration would serve as the "basis for UNICEF policies and actions in support of infant and young child feeding". In May 1996 the World Health Assembly passed a resolution on Infant and Young Child Nutrition (WHA 49.15) confirming its support for the Innocenti Declaration.

• The World Summit for Children held in September 1990 called for "Empowerment of all women to breast-feed their children exclusively for four to six months and to continue breastfeeding, with complementary food, well into the second year."

• In 1992 the *World Declaration and Plan of Action for Nutrition* adopted at the International Conference on Nutrition in Rome, pledged "to reduce substantially within this decade . . . social and other impediments to optimal breastfeeding". The *Plan of Action* asserted, in article 30, "Breastfeeding is the most secure means of assuring the food security of infants and should be promoted and protected through appropriate policies and programmes." Article 33 stated that "Governments, in cooperation with all concerned parties, should . . . prevent food-borne and water-borne diseases and other infections in infants and young children by encouraging and enabling women to breast-feed exclusively during the first four to six months of their children's lives." Article 34 provided a detailed call for action on promoting breastfeeding.

- In 1995 the *Platform for Action* that came out of the Fourth World Conference on Women in Beijing called for promoting public information on the benefits of breastfeeding, implementing the *International Code of Marketing of Breastmilk Substitutes,* and facilitating breastfeeding by working women.

- In 2003 the World Health Organization released its *Global Strategy for Infant and Young Child Feeding*. It recommends that all infants should be exclusively breastfed for six months, and breastfeeding should be continued, with appropriate complementary feeding, for up to two years and beyond (World Health Organization 2003).

A great deal of work remains to be done to clarify the ways in which human rights law and principles should apply in relation to the feeding of infants. The meaning and implications of terms such as "the best interests of the child", "safety", "adequacy", and "the highest attainable standard of health" need to be worked out. This fresh effort should build on work already done on these issues, such as that done on the concept of the highest attainable standard of health (United Nations. Economic and Social Council 2000).

Much of the debate centers on differences in views on breastmilk substitutes such as infant formula. Some people see it as a convenient replacement for breastmilk, while a few see it as close to poison. Others are arrayed somewhere in between. In localities where there is strong evidence and a clear consensus that the use of formula would be seriously dangerous, it would be sensible to adopt rules limiting its use. Until there is broad consensus on this, the best universal rule would be to rely on well-informed choice, with mothers having a clearly recognized right to good information on the risks of using different feeding methods in their local circumstances, combined with suitable support mechanisms to ensure that mothers do have an opportunity to choose (Minchin 2015a). Far too many women feel that with the economic demands placed on them, they cannot realistically choose to breastfeed.

OPTIONAL PROTOCOL ON CHILDREN'S NUTRITION

There have been many authoritative statements at the global level on the nutrition of infant and young children, but they are scattered. The Committee on World Food Security, the apex United Nations group concerned with that issue gives little attention to the food security of children. The annual reports on *The State of Food Insecurity in the World* say little about children, the most vulnerable group. The discourse on right to food gives little attention to the distinct character of the right to food of infants and young children (United Nations. General Assembly. Human Rights Council 2016).

The common tendency to equate food systems with agriculture systems contributes to the tendency to overlook the very different food system for infants. The inattention to the food needs of infants may be partly because commercial interests have a strong influence on the agendas of national and global agencies that deal with food issues.

Many issues relating to children's right to food should be addressed by the global human rights system, especially in the face of the rapid globalization of the baby food industry. One way to deal with them in a comprehensive way would be for the countries of the world to negotiate a new Optional Protocol on Children's Nutrition (OPCN), to be linked to the United Nations Convention on the Rights of the Child (Kent 2015; also see Forsyth 2013; Joint Statement 2016).

The issues can be addressed through a global discussion of the major issues. Areas of agreement could then be spelled out in the OPCN, a statement of broad principles in the human rights framework. Subsequent documents focused on specific issues could set out widely accepted guidelines, perhaps in the form of General Comments or Voluntary Guidelines. These documents could include updates of the Code, the Global Strategy, the Innocenti Declaration and other documents that already reflect widespread global agreement.

Countries that agree to the OPCN's principles by signing and ratifying it could then formulate national legislation that sets out the details of how they understand and plan to carry out their commitments, in the context of their own national human rights systems.

All governments have obligations to respect, protect, facilitate, and provide in relation to the right to food of all people under their jurisdiction. This is well explained in the literature on the right to food (Kent 2005a, 104-110; United Nations. Economic and Social Council 1999, para. 15). When that right is examined in relation to infants and young children, there can be no doubt that large-scale government subsidies of infant formula fail to respect and protect their primary source of food security by undermining breastfeeding. In providing those subsidies, governments violate children's right to food.

CHAPTER SEVEN

HUMAN MILK

Infants should be breastfed by their mothers. Many measures are taken by various agencies to support that. However, when infants are not breastfed by their own mothers, human milk can be supplied to them through milk banking or sharing (American Academy of Pediatrics 2012).

In banking there is no need for contact between the primary providers of the human milk and the infant's caretakers. Sharing arrangements involve direct contact. For example, advertisements may be placed in newspapers or on the Internet to link the human milk providers directly to the infant's mother or other caretakers. Wet nursing is another form of sharing, with no banking stage between the provider and the final consumer of the human milk.

Banking involves collecting human milk at a central place, the milk bank, and then having infants' caretakers obtain milk from the bank. Usually some processing is done at the bank such as pasteurization and quality testing. The women who provide human milk to the banks are likely to be screened through questionnaires and interviews.

Women have provided their breastmilk to other women's infants since time immemorial (Baumslag and Michels 1995; Golden 2001; Palmer 2009, 182-190, Swanson 2014). Sharing of breastmilk is currently underway and growing in many countries (Akré 2012; Akré, Gribble, and Minchin 2011). To illustrate, there is a milk sharing organization called Human Milk 4 Human Babies Global Network (hm4hb 2014). It promotes sharing of breastmilk freely, with no compensation to the providers. Their mission is "to promote the nourishment of babies and children around the world with human milk." Another organization called Only the Breast functions as "A community for moms to buy, sell, & donate natural breast milk (Only the Breast 2016)" It arranges classified advertisements to connect sellers or donors to buyers. It exercises little control over the transactions.

There have been advances in methods of delivery of human milk to infants. Some women who are unable to breastfeed directly in the conventional way can express their milk and deliver it through a tube or bottle. Or they can deliver milk obtained from another woman. The human milk can be fed to the infant by the mother, the father, or another caretaker. In many cases, direct skin-to-skin contact is arranged.

Banking usually is for critically ill infants, those who are in life-threatening conditions, especially those in hospitals. Sharing usually is for infants who are not critically ill. Milk banking is about saving lives while sharing is about improving infants' health.

Arrangements for sharing and banking human milk have become more systematic, raising many issues (Arnold 2006; Aunpalmquist 2014; Dutton 2011; Fentiman 2010; Gribble and Hausman 2012; Heise 2014; Levine 2016; Palmquist and Doehler 2014; United States Food and Drug Administration 2015). The issues are very different from those raised in dealing with breastmilk substitutes such as commercial infant formula (Brady 2012).

This chapter focuses on milk banking or sharing to provide breastmilk for normal infants who do not get breastmilk from their own mothers This is not about the special issues that arise when collecting breastmilk for research purposes or to provide breastmilk to critically ill infants such as those in neonatal intensive care units.

Milk banks could benefit many infants beyond those who are critically ill, and do this in ways that strengthen supplies for those who are critically ill, Expanding the reach of human milk banks could benefit many infants who otherwise would be fed with infant formula.

Feeding infants with human milk from milk banks is not as good for them as being breastfed by their own mothers, for several reasons.

- The milk is not as fresh as that obtained through direct breastfeeding.

- Banked milk might not be carefully matched for the age of the child.

- Banked milk cannot change in response to immediate short-term needs such as those associated with infections or the time of day (Bologna 2016; Petherick 2016)

- Pasteurization, freezing, and other processing are likely to lead to some deterioration of the milk.

- Banked milk is often fed through bottles, with little direct skin-to-skin contact.

Breastmilk's role in strengthening the infant's immune system can occur even if the milk is from a woman other than the biological mother. In some circumstances, it can be advantageous to get other women's milk. Dr. Gro Nylander illustrated this with the case of a child in Norway with an antibiotic resistant infection. By sourcing donor milk from a mother in an immigrant community in which the disease was more common, she established the immunity needed to overcome the infection (Minchin 2016).

In another context, Professor Ameae Walker felt the use of wet-nurses was beneficial for similar reasons:

> Wet nurses were usually from a lower social echelon, and were likely exposed to more infectious agents. They had to have very robust immune systems to survive, and transfer of antibodies and cells through their milk probably contributed greatly to the survival of royal babies. What the research is telling us is that a baby would likely gain from nursing by multiple women with different immune backgrounds. Not too long ago in most cultures, if a baby cried a woman would pick it up and nurse it; most women were pregnant or lactating. Now, socially, we would think that rather peculiar -- to nurse someone else's baby. But is the change in how society thinks about this issue the best for the baby's health? (University of California—Riverside. 2016)

This might give new meaning to the adage, "it takes a village to raise a child." In so-called primitive societies, where infants and young children freely move from one woman to another for feeding and care, they might do better than they would living in sterile environments with overly protective mothers

Studies are needed to determine the difference in health impacts of feeding through direct breastfeeding and banked milk, and between different methods of processing banked milk and managing milk sharing (e.g., Medical Xpress 2015; Salcedo et al., 2015). As methods for quality assessment are improved, it might be possible to devise procedures that assess breastmilk, banked human milk, and infant formula by the same standards, and thus support comparisons among them (Johannes 2015). Human milk analyzers like that offered by Miris in Sweden might be improved to cover more important dimensions (Miris 2016).

It is reasonable to accept that feeding children with human milk from milk banks is not as good as direct breastfeeding by the biological mother. It is also reasonable to accept that feeding with banked milk, if well-regulated and well-managed, is significantly superior to any other alternatives to direct breastfeeding. As argued in Chapter One, it is feeding with human milk, not formula, that is the best alternative to direct breastfeeding by the biological mother.

Some things fed to infants are worse than formula, such as feeding with unmodified cow or goat milk, evaporated milk, condensed milk, cola, tea, honey and ghee, or herbal concoctions. Such practices should be replaced by direct breastfeeding by the biological mother wherever possible, but if that cannot be done, providing human milk is likely to be better than feeding with formula.

Making banked human milk readily available would be a way of pushing back against pressure from the formula manufacturers (Schiller 2016). This is important because the manufacturers give little serious attention to the impact infant formula is likely to have on the health of children and the adults they will become.

COMPETE OR COLLABORATE?

Some resist the idea of offering banked milk to children who are not critically ill based on the principle that those who are sickest should be the highest priority (Wardlaw et al. 2014). It is important to save the lives of those who face immediate risk of death. This position is understandable, but the issue is not so simple.

First, there are triage considerations. Where there are limited health care resources, giving a great deal of attention to the worst cases might mean others would be deprived of care. Specialists in neonatology have made great advances in saving the lives of infants who are born prematurely (Belluck 2015). However, as they push toward saving infants who are born more and more prematurely, the costs for each save goes up. It is important to find a reasonable balance between serving extremely needy patients and serving those who are less needy.

Second, there might be mistaken assumptions about limitations in supply. In some settings, the supply of human milk is extremely limited, as in those milk banks in which the only providers are the mothers of hospitalized infants. However, many more women would be willing providers if they were well informed about the opportunity, and even more if extra incentives were offered. More women might offer their services if they were assured of a pleasant experience and perhaps given recognition and small gifts as tokens of appreciation.

Still more women might offer their milk if they were paid directly for it. The supply could be expanded further by reaching out to women who were averse to coming into hospitals, but would gladly come to a human milk collection center at, say, a shopping mall, along with a few friends. In several countries, there are now arrangements under which women can express their milk at home and then drop it off and nearby collection points.

Some critics worry that milk banks that pay their providers could reduce the supply available to banks that that do not pay their providers (Buia 2015). However, the overall supply of banked milk could be increased with appropriate inducements. The major increases would come not so much from increasing the production of individual women but rather by drawing in more women as providers. The potential supply is far greater than the need.

Third, it should not be assumed that milk banking for critically ill infants must be in direct competition with milk banking for infants who are not critically ill. With suitable arrangements, the two might complement one another.

To illustrate, large milk banks could sell their pasteurized and bottled human milk at different prices to different categories of customers. Earnings from providing human milk for infants outside hospitals could be used to subsidize supplies for those in hospitals. Instead of simply rejecting, say, adult body builders and ice cream makers, the bank could charge them high prices. Cross-subsidies between different categories of customers could be set up, with less needy users charged higher prices so that more needy users could be charged lower prices. This pricing strategy was applied in early milk selling practices in the United States where, "The hope was to have enough families paying well over 17 cents an ounce to subsidize those families that paid nothing (Swanson 2014, 38, also see 250)."

Milk banks that serve infants who are not in hospitals could absorb some of the processing costs for human milk destined for use in hospitals, perhaps by allowing hospital milk banks to use their equipment. There might be economies of scale under which both groups could benefit.

Milk banks for critically ill children could be set up separately from those for children who are not critically ill, or they could be combined. Either way, competition between them could be replaced by collaboration.

Well-managed human milk banks could play a significant role in breastfeeding promotion:

> When supported by national and regional government engagement in a comprehensive push for prioritizing breastfeeding promotion, an HMB can serve as an integral component of a community breastfeeding program by providing lactation promotion, infant feeding support, and education on maternal and infant health. The existence of HMBs in communities helps to increase breastfeeding rates by highlighting the value of breast milk and the importance of early and exclusive feeding of human milk. Increased community knowledge about the importance of breast milk builds a larger donor pool, establishing a solid foundation for HMBs to ensure that safe donor milk is available. (PATH 2013, 68)

As one infant feeding expert argues, "greater trading and exchange of human milk could mean greater societal recognition of the economic value of breastfeeding, and might enable human milk or breastfeeding to better compete with bovine milk-based substitutes (Smith 2015)."

Using milk banks for infants who are not critically ill would not necessarily lead to a reduction in supply for those who are critically ill. Serving both groups simultaneously could benefit both.

INCENTIVES FOR PROVIDERS

Eats on Feets is devoted to community breastmilk sharing. It says:

> Community breastmilk sharing works because mothers, fathers, professionals, communities, caring citizens and people just like YOU are joining together to help ensure that babies have access to commerce-free breastmilk.

> Babies need breastmilk to maintain optimum health. Parents and professionals know this! Every day, women from around the world selflessly donate thousands of ounces of breastmilk directly to babies. With Eats On Feets, these donations are commerce-free, just as nature intended, and they are making a huge difference in the lives of babies and their families. (Eats on Feets 2015)

Their system revives and extends the practice of wet nursing. But it is not clear why Eats on Feets feels that sharing must be "commerce-free, just as nature intended." It is good that it could deliver thousands of ounces, but if the providers were compensated, it might be able to deliver millions of ounces.

Women who supply their milk to banks are commonly described as *donors*. However, there are some human milk banks that pay them directly with cash, or they may be compensated in other ways. They are all described here as milk *providers*, whether or not they are paid.

There are sharp divisions on the idea of compensating providers. For example, the Human Milk Banking Association of North America opposes payment, saying, "accepting milk donations from volunteer donors is the most ethical way to ensure that milk donations will be shared with the most critically ill of infants." It argues, "Through following the nonprofit model of milk banking, HMBANA milk banks prioritize infant health when distributing donor milk to fragile infants. When donors contribute their milk to a for-profit breast milk operation, they do not always have the level of certainty about the destination of their donation (PRWeb 2014)."

HMBANA said, "By donating their milk with a nonprofit milk bank, mothers can be sure that their milk will be allocated to the sickest babies in their area." It is not clear what non-profit status has to do with the fact that human milk providers are not compensated. Also, it is not clear what non-profit status has to do with the credibility of their assurances regarding which infants will get the milk. The relevant laws regarding nonprofit corporations vary a great deal across different jurisdictions, so it is difficult to generalize. In the United States, a non-profit corporation is one in which stockholders are not allowed to make a profit. They can pay for goods and services provided to them. They can pay their employees.

HMBANA suggests that non-profit organizations are inherently more ethical than for-profit organizations, but it offers no argument or evidence to support that idea. An organization that is nominally non-profit can be exploitative, perhaps by charging hospitals high prices for milk they receive from women who are not compensated at all. It could be just as exploitative as organizations that are organized on a for-profit basis.

One view in Australia is that:

> A prohibition on payment protects both the donor and the recipient: it can avoid inducing donors to compromise their (or their babies') health by giving too much and it protects recipients from the risk that unhealthy donors may have been attracted by the prospect of payment. (Commonwealth of Australia 2014; Sansom 2015)

The European Milk Bank Association said:

> Milk banks in Europe do not profit from or commercialise the provision of human milk. EMBA believes that the sharing of human milk is a humanitarian and altruistic act and that the provision of human milk

should always be without commercial aspects. Every initiative which involves any form of payment (other than reimbursement to mothers of their expenses) or business with human milk should be considered unethical and proscribed. (EMBA 2011)

This is too sweeping a rejection of the idea of payment. Prohibiting payment to the providers of human milk can limit the supply, with the result that it cannot be offered to many infants who might benefit from it. Thus, rejecting the idea of payment is ethically questionable. Commerce certainly can lead to abuses (Fentiman 2010), but it is the abuses that should be abolished, not commerce itself. The risks associated with paying women for their milk might be manageable.

An opinion piece in the *New York Times* agreed the benefits of human milk are not just for sick or premature infants. The author acknowledged, "A market for breast milk seems like the logical solution for matching the deluge of milk some women produce to the desperate need for milk that some babies and hospitals have (Currid-Halkett 2015)." However, the author then complained that some companies use human milk to manufacture products they sell at high prices, and sets that as the basis for arguing that women should donate rather than sell their human milk. How does that follow? If the supply of human milk for infants who are not critically ill could be greatly increased by allowing women to sell it, that option should be retained and strengthened. If the manufacturers of those expensive products should be regulated in some way, that problem should be addressed directly.

Gabrielle Palmer argues that women should not be paid for their human milk because "altruism is important to maintain quality":

Human milk donation requires trust between donor and receiver. If a mother expresses her milk for money, then she may be tempted not to tell the milk bank that, for example, she occasionally smokes or that she has had a course of antibiotics. Just as with blood or organ donation, financial incentives can do harm. (Palmer 2009, 330)

Similarly, paid providers might be tempted to add cow's milk or something else to their breastmilk to increase the volume and thus increase the payments they receive. There are ways to control such abuses that are likely to be more effective than barring compensation to providers. For example, the milk that is provided could be tested for contaminants.

There is a long history of paying wet nurses who breastfeed their client infants directly and also women who provide their breastmilk in sterilized bottles. "In 1929 at least twenty American cities had a mothers' milk station, buying and selling human milk (Swanson 2014, 33)."

The line between paying and not paying women for their milk is not sharp and clear. Instead of cash, incentives of various forms could be offered, such as small gifts, recognition, rides to and from the collection site, and hospitality at the collection site. Food could be made available for consumption at the site and to take home. Lactogenic foods or "galactogogues" -- foods that support milk production--would be especially appropriate (Jacobson 2015; Murray 2015).

Women who provide their milk to other families could be compensated whether or not the milk is destined for infants who are critically ill. This could be done regardless of whether it was handled by a for-profit or non-profit organization. Both types should be closely regulated and monitored by government and by relevant professional associations. There are several key points:

- It would be fair to compensate women who provide human milk for others' infants.

- With compensation, there would likely be a larger supply, so more infants would be able to receive breastmilk rather than breastmilk substitutes.

- With a larger supply, the cost for human milk would be lower.

- With more children getting human milk rather than formula or other breastmilk substitutes, the health and well-being of both women and children would be substantially improved.

- The money that now goes to infant formula manufacturers would be better spent if it went to women, especially women with low incomes.

Milk banking organizations that wish to operate solely with voluntary donors should be free to continue that way. At the same time, other organizations should be allowed to compensate women who provide their milk. The providers don't all have to be donors.

In some cases, the cost of human milk might be covered by health insurance programs, but that coverage is uneven. In one situation, the mother of a hospitalized premature infant found she was unable to produce enough to keep up with her infant's needs. The hospital told her that, because her health insurance would not cover the cost, she would have to pay US$250 a day for milk from the New York Milk Bank. She was told that as the infant got older the cost might increase to US$500 a day (Formoso 2016).

It is not clear how those high daily rates could be explained, especially if the milk bank does not pay its providers. What is clear is that, at even lower payment rates, many women would be glad to deliver the product in person, on demand, in a form that is safe and nutritionally superior to any infant formula that could be offered.

Consideration should be given to possibilities for reviving wet nursing under some conditions, as an alternative to milk banks. As Maureen Minchin put it to me:

> Parents should be able to make such arrangements for the sake of their child's health, and pediatricians and nurses should respect parents' choices, while advising on necessary basic hygiene precautions as needed. Health professionals may need some re-education to overcome cultural prejudices arising from a lack of knowledge of relative risks, and to adopt an evidence-based support strategy.

There is a need for creativity not only in the ways in which human milk is collected, but also in the ways in which it is distributed. Much can be learned from the Philippines:

> The Mandaluyong City government has renewed its milk-letting program with the Department of (DOH), wherein volunteer mothers can donate breast milk to the Philippine General Hospital's (PGH) Department of Pediatrics. . . . Mothers who join the city's Breastfeeding Patrol do not only donate breast milk, which is collected by and stored in the milk bank of the Mandaluyong City Medical Center. When major disasters hit the city, the mothers are also dispatched to evacuation shelters so they can provide milk for infants . . . . As part of the city's nutrition program, the Breastfeeding Patrol has been successful in helping mothers who have

a hard time producing their own milk. "Parents should be practical because milk formulas are very expensive. And for mothers who cannot produce milk, the Breast-feeding Patrol is our answer," the mayor said in a statement. (Metro Briefs 2016)

It is not difficult to imagine Breastfeeding Patrols in many different communities, attending to emergencies and other kinds of urgent needs at the community level and at the family level. One can also imagine Breastfeeding Patrols operating internationally. In 2009 actress Salma Hayek got widespread media attention by breastfeeding a hungry infant in Sierra Leone, the country with the highest child mortality in the world (Kaplan 2009). With the right arrangements, many lactating women might be willing to do something similar for several weeks, perhaps for needy infants who are nearby.

Some creativity in the collection and distribution of human milk does not make much sense. Why would one want to deliver 2,300 pounds of human milk from California to South Africa for any reason other than to show off the cold chain logistic transport capabilities of a shipping company (Vigliarolo 2016)? Surely it would cost much less to collect milk from women in South Africa, and it would be kinder to the environment.

REGULATION

There are risks related to human milk banking, but relying on altruism through non-payment is not a good way to deal with them. The risks can be reduced with appropriate monitoring and regulations, including screening of providers. No one asks farmers or infant formula manufacturers to offer their products for free as a way of ensuring their integrity. Why should women who offer human milk for sale be treated differently?

The late Miriam Labbok, a well-known breastfeeding expert, said "offering payment may seem generous . . . or coercive" and added:

> The heart of the issue to me is one of availability [of] full unbiased information and free choice among choices in a system that is free of fiscal or personal or health system coercion. Unfortunately, ours is not such a system. (Labbok 2015)

Avoiding coercion certainly is important. However, telling women that they must not accept payment for their milk is itself a form of coercion.

Labbok also said, "When we ask for women to sell their milk when their child is still nursing, we are asking that the milk, even if there is surplus, be denied to their own child (Labbok 2015)." This could happen, and it would be a serious matter. However, Pamela Morrison, another breastfeeding expert, told me:

> There could be checks and balances like clinics attached to milk banks attesting to good gain etc. and the nature of human milk synthesis being what it is, most women can easily increase their supply to make an ounce or two extra every day over and above what their baby needs - heavens, most women can easily breastfeed twins.

Some women who are paid for their milk might for that reason deprive their own infants of it. Several measures could be taken to limit that risk:

- Issue clear instructions to providers regarding the breast-feeding of their own infants.

- Checks could be made to determine whether the women are breastfeeding their own infants by checking the infants' stools.

- Monitor the health of their infants through visits, by obtaining reports from their health care providers, or through direct examination by pediatric health care workers.

- Ensure that the price offered for human milk is not so high as to create a strong economic incentive for women to sell their own milk and use breastmilk substitutes to feed their own infants.

Measures of this sort were used in a milk bank in Leipzig, Germany. In 1989, 95 paid donors supplied 10,000 liters of breastmilk while still meeting the needs of their own infants. Some breastfed longer (Minchin, 2015b). The bank would not accept women's breastmilk unless they continued to breastfeed their own infants.

In some cases, the financial incentive to mothers might lead them to breastfeed their own infants for a longer time than they would otherwise, as they did in Leipzig (Wells 2015). Whether paying providers would in fact deter many mothers from breastfeeding their own infants is an empirical question, one that should be closely monitored.

If women are offered money for their milk, and they can get free or subsidized infant formula, some might be tempted to get that formula to feed their own infants, and pump their own milk to sell it (Rinker 2016). The sensible way to address this problem is not to refuse to pay for women's milk, but to limit the distribution of free or subsidized formula.

There are now well developed procedures and guidelines for addressing safety and other potential problems in milk banking, reducing the risks (NICE 2010; PATH 2013, 2015). The systems are still imperfect, but there is steady progress in the articulation of good practices in the sharing and banking of human milk.

Laws in the United States relating to breastfeeding are summarized in a website of the National Conference of State Legislatures, accessible at http://www.ncsl.org/research/health/breastfeeding-state-laws.aspx Aparrently, there is no place in the U.S. where the selling of breastmilk is illegal. In some contexts, people interpret certain laws as meaning that the selling of breastmilk is illegal while others interpret the same law differently. There is a need for new, clearer, wiser law relating to the buying and selling of breastmilk under various conditions (Dawson 2011), not just in the U.S., but everywhere.

The supply of human milk could be increased with appropriately designed incentives for human milk providers. Many women and many infants could benefit. These are benefits that should not be foregone without good reasons. There is nothing intrinsically wrong with paying women who provide their breastmilk for infants not their own.

WOMEN'S DIGNITY

Some people object to the idea of women selling their milk (Carter, Reyes-Foster, and Rogers 2015). They feel "treating them like cows" would violate their dignity. Others disagree, arguing that if it is managed well, providing human milk to banks could be empowering for women and strengthen their roles in society. Instead of debating whether human milk should be treated as gift or as a commodity, there are ways to value it and treat it as both.

Much can be learned by comparing the treatment of human milk with the treatment of other body products such as blood, sperm, eggs, and organs (Lewin 2015; Swanson 2014).

Women, human milk and breastfeeding have been consistently undervalued, not given the recognition they deserve for their contribution to human well-being. Several writers have argued that the value of human milk should be fully recognized in national accounts and in indicators such as the gross national product, and women should be recognized for its production (Aguayo and Ross 2002; Berg 1973; Hatløy and Oshaug 1997; Oshaug and Botten 1993; Palmer 2009, 319-344; Rohde 1982). In Australia, for example, human milk production is estimated to be worth more than $3 billion a year (Smith 2013).

Breastmilk is valuable, but usually the producers are not compensated for it (Palmer 2009, 331-339; Smith, Galtry, and Salmon 2014). Some have argued that women should be paid for breastfeeding their own infants and other services they normally provide without compensation (Francis et al. 2002). That may not be feasible, but women can be and often are paid when they provide human milk for infants not their own.

Some people seem to think having women carry heavy loads at construction sites in exchange for money is acceptable, but paying them for what they are uniquely equipped to do is not. Why is it that women are so often called on to volunteer their services, while men expect to be paid quite handsomely for theirs? Human milk is being bought and sold. Following Julie Smith, we should ask, "How can we improve on the present situation where everyone except the woman who donates her milk benefits? (Smith 2015)."

Anyone's deciding for women as a group what they should or should not do would itself be a violation of their dignity. It would be coercive. No one is proposing that women should be forced into providing their milk to milk banks. Where the conditions are well managed, women who wish to offer their milk, whether for free or in exchange for compensation of some sort, should not be prevented from doing that.

HUMAN MILK BANKING
AS SOCIAL BUSINESS

As early as 1925, the Journal of the American Medical Association editorialized about the prospects for dried human milk. The efforts seemed entirely feasible, but JAMA judged that the efforts "cannot be said to have reached a dividend paying basis, unless one is content to count as such dividend the life-saving quality this commodity is known to possess when given to certain sick or premature infants (JAMA 1925; also see Smith, L. W. 1924)."

We can only speculate about how the history of infant feeding might have been different if the value of that life-saving had been factored into the calculations. The feasibility of doing that is clear when consideration is given to covering human milk in health insurance programs (Campbell 2016). If it makes sense for some governments to provide free or highly subsidized infant formula, surely it would make even more sense for governments to pay a share of the costs for human milk.

There are ways to deliver social benefits in a business-like manner through what has been called social business (Yunus 2007). At the very least, "business-like" means revenues cover costs so the activity can be sustained. In many places, there is also the option of creating a for-profit Benefit Corporation (Benefit Corporation 2016). Non-profit corporations can be designed to produce social benefits, in accordance with local legal frameworks.

With these business models, groups of entrepreneurial breastfeeding advocates could start milk bank businesses whose primary purpose is improving the health of infants. Money would have to be earned to cover the expenses of the operation. That would include providing suitable incentives to the women who provide the human milk and paying all the workers who operate the organization.

Enterprises of this sort are already starting up. Currently most milk banks focus on meeting the needs of critically ill infants, but there is growing interest in providing human milk to infants who are not critically ill. Medolac Laboratories offers "Commercially Sterile, Shelf-Life-Stable Human Donor Milk as Easy-to-Use as Formula (Medolac Laboratories 2015)." Another new powdered human milk product promises "The convenience of formula" along with "The health benefits of breast milk" (Mammilla 2015)." They offer up to one dollar per ounce to women who provide their milk. The packaged product sells for between two and three US dollars per ounce, depending on the quantity purchased. One company offers milk from providers in Cambodia to customers in the U.S. (Ambrosia Labs 2016; Springwise 2016 Wood 2015).

These early entries into the field are not well regulated, and some are highly questionable. The Ambrosia Labs website is incorrect, where it states, under Frequently Asked Questions, that "Breast milk is regulated by the Food and Drug Administration (FDA) as a food item."

The plain human milk that is offered for sale should not be confused with the high-priced fortified human milk intended for infants who are ill and need special treatment in hospital settings.

There are serious questions relating to the value of these fortified products in treating infants who are ill, such as possibly misleading claims about the benefits they can provide (Rinker 2016). There are serious concerns about potential abuses by the companies that produce those highly specialized human milk products. However, those abuses are not a sound basis for arguing that no women should be paid for the milk they provide.

Businesses designed to make human milk more readily available could be set up in low-income as well as high-income countries. There would be a variety of problems to overcome, as in any business startup. If the managers of the project were mainly interested in the "bottom line" of improving children's health rather than the bottom line of taking home a lot of money, these businesses could lead to substantial benefits for both women and children.

Where such policies would be useful, it could be required that these enterprises are managed entirely by women. Men could help in the start-up phase, but following that start-up period, men could be limited to advisory roles, as determined by the managers.

These enterprises could begin as small pilot projects, with well-chosen advisors who would seek ways to allay fears about safety and other concerns. Appropriate bodies of government could develop systems for regulating these enterprises. Transparency in financial arrangements could help to limit abuses.

Such initiatives could lead to the creation of many small businesses. Some would succeed and some would fail, and a great deal would be learned in the process. National governments could establish central offices to oversee them. Global agencies such as UNICEF and the World Health Organization could lead the effort to draw up suitable guidelines for managing human milk banks.

In some places, it might not be possible for milk banks to operate as self-sufficient and economically sustainable businesses. They might be helped with direct subsidies from governments, gifts from private parties, or various kinds of creative arrangements. As suggested earlier, some hospitals might provide space for them to function. Though subsidized, they could still operate in a business-like way, and therefore need less support from outsiders than they would as purely voluntary organizations. The scale of their operations would have more room to grow if they operated as businesses rather than wholly voluntary operations.

Feeding infants with banked milk is likely to be more expensive than feeding with formula. However, over time, the cost of banked milk will be reduced as volumes increase and efficiency in handling is improved, so it could become more accessible. The demand is there, and many people and insurance agencies might be willing and able to pay for a high-priced product (Bye 2016) because the health of infants is at stake.

For people with low incomes, infant formula remains out of reach and may be difficult to use because of sanitation and other issues. In those cases, it is especially important to promote improved breastfeeding practices so that mothers can feed their own children more effectively. Where that fails, increasing the supply of breastmilk from other women could do a great deal to improve children's health status, sharply improving their survival prospects (Patel 2014; UNICEF 2013).

Maureen Minchin asks why so much effort goes into the development of infant formulas:

> Why is any of this vast scientific enterprise necessary, when women make a far better product that poses far fewer risks to their infants, and provides major benefits

for both women themselves and their children? Human milk could be much more widely available than it currently is, and even be used as medicine, not only for infants. (Minchin 2016)

Gabrielle Palmer said this another way:

> If a multinational company developed a product that was a nutritionally balanced and delicious food, a wonder drug that both prevented and treated disease, cost almost nothing to produce and could be delivered in quantities controlled by the consumers' needs, the very announcement of their find would send their shares rocketing to the top of the stock market. The scientists who developed the product would win prizes and the wealth and influence of everyone involved would increase dramatically. Women have been producing such a miraculous substance, breastmilk, since the beginning of human existence. (Palmer 2009)

The subsidization of infant formula by governments means they recognize that protecting the health of infants is a public good. The funding presently devoted to purchasing formula could be redirected to better alternatives. National governments and the international community should provide greater support for optimum breastfeeding (Holla et al. 2013), and should also subsidize human milk for infants who are not breastfed by their own mothers. If the governments of large countries such as China, India, and the United States vigorously promoted optimal breastfeeding, and also subsidized human milk banking and sharing, their practices could have huge health benefits within their own countries, and serve as important models for other countries.

Why should human milk banks and milk sharing arrangements be created for critically ill infants and not for the many other children who could benefit from them? There are risks, but they could be monitored and managed with sensible regulation. Where mothers do not breastfeed their own infants, increasing the supply of human milk from other women could do a great deal to improve infants' short-term and long-term health, and the health of their mothers as well.

CHAPTER EIGHT

THE INADEQUACY OF INFANT FORMULA

This discussion of the inadequacy of feeding with infant formula has been saved for last because of its importance. Governments would not subsidize or otherwise promote feeding with infant formula if they appreciated how deficient it is when compared with breastfeeding.

The basic scientific point, that all formula is inadequate, was made in the opening chapter of this book, where it was established that formula regularly leads to worse health outcomes than breastfeeding. As argued repeatedly in the preceding chapters, to achieve the best possible health for infants, mothers should be provided with every possible support for breastfeeding, at the family level and at the societal level.

Chapter One said all alternative feeding methods are worse for infants' health than breastfeeding. Here we ask, why are they worse?

MORE THAN THE SUM OF ITS PARTS

As explained in Chapter One, the ingredients list for infant formula is set out in global recommendations from the Codex Alimentarius Commission. In 1976 the Commission issued a *Statement on*

Infant Formula that said, "Numerous formulae have been produced which offer a nutritionally adequate food for infants, and, provided they are prepared under hygienic conditions and given in adequate amounts, there is no contra-indication to the use of such products (Codex Alimentarius Commission 1976)."

No contra-indication? That suggests that, in relation to infants' health, feeding with nutritionally adequate infant formula is as good as breastfeeding. If formula was nutritionally adequate, there would be no difference in expected health outcomes between the two methods of feeding.

However, elsewhere, in relation to other foods, the Codex Alimentarius Commission said:

> The nutritional adequacy of a product can be defined in terms of protein quality and quantity and content of minerals and vitamins.
> Such a product should be considered nutritionally equivalent if:
> (i) its protein quality is not less than that of the original product or is equivalent to that of casein and
> (ii) it contains the equivalent quantity of protein (N _ 6.25) and those vitamins and minerals which are present in significant amounts in the original animal products. (Codex Alimentarius Commission 1989, Section 7.2)

When applied to infant formula, presumably equivalence refers to comparisons with breastmilk. In this approach, if the ingredients in infant formula match the ingredients in breastmilk, then the formula would be described as nutritionally adequate.

Thus, we have two different and incompatible understandings of the meaning of adequacy, one centered on health outcomes, and one centered on matching lists of ingredients. That is a huge difference.

Within the Codex framework, various countries and regional groups set out their own variations in the formula ingredients list, based on the allowance for variations in the Codex guidelines. In the United States:

> The Infant Formula Act of 1980 authorized the Food and Drug Administration (FDA) to assure quality control of infant formulas . . . . Based on the recommendations of the AAP [American Academy of Pediatrics] , the FDA requires the following nutrients be present in all infant formulas: protein; fat; vitamins C, A, D, E, K, B1, B2, B6, and B12; niacin; folic acid; pantothenic acid; calcium; phosphorous; magnesium; iron; zinc; manganese; copper; iodine; sodium; potassium; and chloride. (Stevens, Patrick, and Pickler 2009, 37)

This focus on ingredients is common to all high-income countries. Many other countries follow their lead.

BEYOND NUTRIENTS

Since 1976 it has been the ingredients focus that has dominated worldwide, not the health outcomes focus. The design of infant formula or entire diets has been based on the idea of getting the intake of ingredients right (American Pregnancy Association 2016; Institute of Medicine 2004; The Milk Meg 2013; Ralston 2006).

This reductionist approach focusing on ingredients is troublesome in dealing with food and nutrition generally, and not just for infant formula. As nutritionist Carlos Monteiro explains:

> Nutrition science is taught and practiced as a biochemical discipline. Practically all nutritionists now categorise food in terms of its chemical composition, as do most lay writers. This almost universal perception of nutrition is evident in textbooks and scientific journals, and on food labels, journalism, and 'diet books'. The identification of food with its chemistry is a defining characteristic of modern nutrition science, as invented in the early 19th century. Seeing food in terms of its chemistry has enabled the industrialisation of food systems. In particular, it has made possible the formulation of ultra-processed products from 'refined' or 'purified' chemical constituents of foods – oils, proteins, carbohydrates, and their fractions – together with 'micronutrients' – vitamins and minerals. (Monteiro 2011)

He summarizes: "Identification of food mainly with its chemical constituents at best has limited value, and in general has proved to be unhelpful, misleading, and harmful to public health."

The reasoning behind the ingredients focus is faulty. It is like saying that to get a Rembrandt painting, all you need is some red, blue, and yellow paint.

It is like saying, if you put bits of some sort of metal, some sort of plastic, and some sort of glass into a bag and shake it up, you can expect the resulting mix to act like a car.

It is like saying that you could collect the ingredients for a gourmet meal, mix it in a blender, and serve it up in a high-end restaurant.

A soil scientist described the folly of simply adding nitrogen, phosphorus, and potassium to soil, saying, "The idea that we could ever substitute a few solid elements for a whole living system is like thinking an intravenous needle could administer a delicious meal (Barber 2015, 90)."

Like healthy soil, breastmilk should be viewed as a living thing, one that changes dynamically as the infants' needs change. The difference is suggested by the picture on the front cover of this book. Human milk, on the left, shows a variety of forms, dominated by fat globules. Infant formula, on the right, looks like an industrial product, because it is. Things like the immunological factors in breastmilk simply cannot be bottled (Ghosh et al. 2016; Minchin 1998a, 1998b, 2015a, 2015b, 2016). As Maureen Minchin put it, "Breastmilk is alive, a living tissue, 'white blood'. Infant formula is a dehydrated unsterile soup of animal, vegetable, fungal and algal ingredients."

Breastmilk contains many ingredients beyond nutrients, including hormones, bioactive molecules, microorganisms of various kinds, oligosaccharides, and other complex factors still unknown. Even if the ingredients list could somehow match the list for breastmilk (it cannot), that would not mean much. Infant formula should not be evaluated based on its ingredients. The question is, does the concoction *function* in the same way as the product you are trying to emulate. Does it do what it is supposed to do?

FUNCTIONALITY

What is formula supposed to do?

In the basic Codex Standards for Infant Formula, the ingredients are those "which have been proved to be suitable for infant feeding (Codex 2007, Section 3.1.1)." In the Codex Standards for Follow-up Formula, infant is defined as "a person of not more than 12 months of age," and follow-up formula is defined as foods "which have been proved to be suitable for infants from the 6th month on and for young children (Codex Alimentarius Commission 1987)."

How a food could be "proved to be suitable" is not explained in either document. Proven how, and by what agency? In legal terms, "proved" refers to a positive obligation. Under these standards, since no appropriate proof has been presented, those products should not be marketed.

Section 3 of the 2007 document is on Essential Composition and Quality Factors. Composition refers to ingredients, nutrients. What are the Quality Factors? There is no explanation. There is no discussion of any kind about what infant formula is supposed to do, apart from having been proved suitable.

The composition of infant formula has improved over the years, so it now appears to do a reasonably good job. But that thought must be accompanied by serious reservations. Maureen Minchin points out that although formula can keep infants alive . . .

> Some die unnecessarily in every country because of it. Formula-fed babies do grow: so much so that industry has steadily lowered the protein content in an effort to reduce obesity. The FDA has declared that formula feeding is safe enough - even though the powder cannot be sterile and every summer sees gastroenteritis and every winter bronchiolitis among the artificially fed.

In all conditions, at the population level, health outcomes for formula fed infants are worse than those for breastfed infants. Mothers should be supported in breastfeeding, not feeding with formula or with other products that are worse than formula.

Health outcome are worse with formula feeding than with breastfeeding even when both are as safe as possible. Since so much attention has been given to the safety of infant formula, some people might assume that if it is safe, feeding with formula is as good as breastfeeding. However, just as with pharmaceuticals, infant formula should be not only safe but also effective in doing what it is supposed to do. Children should be fed in ways that result in good physical growth, strong immune systems, good visual acuity, good intellectual development, and so on. The method of feeding should limit the likelihood of a broad range of health problems in the infant and in the mother as well.

The things that breastfeeding does are the things that feeding with formula ought to do. Using the Codex language, feeding with infant formula would be "proved to be suitable for infant feeding" when it was shown to regularly result in health outcomes as good as those obtained with breastfeeding. Hardly anyone thinks that is possible.

In the United States, the Food and Drug Administration is the primary agency responsible for ensuring the quality of infant formula. Remarkably, beyond the ingredients list, the only Quality Factors that concern the FDA are normal physical growth and the biological quality of the formula's protein component. In 2014, as they were preparing revisions of the rules, I commented on this to the FDA:

> This means that of the many different functional requirements, the only one to be assessed for infant formula is its efficacy in leading to adequate physical growth in the short term. The language of the rules implies that if an infant formula leads to adequate physical growth over a period as short as fifteen weeks, it is of good quality.
>
> It should not be suggested that quality on a single dimension is sufficient when infant formula must perform well on many different dimensions. There are many studies that demonstrate this. To illustrate, in 21 Dangers of Infant Formula, the World Alliance for

> Breastfeeding Action shows 21 different ways in which feeding with infant formula appears to function less effectively than breastfeeding (see http://www.waba.org.my/whatwedo/advocacy/pdf/21dangers.pdf). Each of them represents a concern about the quality of infant formula.
>
> It is misleading to suggest that a short-term measure of infants' physical growth can reasonably be viewed as a measure of the overall quality of infant formula. (Kent 2014c)

The FDA ignored the point.

We know feeding with infant formula is inadequate because it regularly leads to worse health outcomes. The only exceptions are in cases of rare metabolic diseases of the infant such as galactosemia and phenylketonuria. There are cases in which mothers should not or cannot breastfeed, but that does not mean the infants should not be fed with human milk. Wet nursing and milk banking have roles to play.

The deficiencies of formula cannot be resolved simply by adding new ingredients into the mix. Manufacturers often add a new ingredient and then say the new version of their product is *closer* to breastmilk. That is not the same as saying it is *close* to breastmilk. New York is closer than New Jersey to Paris, but that does not mean New York is close to Paris.

ADEQUACY IS NOT JUST ABOUT NUTRITION

There are several ways of viewing the major functions of breastfeeding. In the United States, the government relies on scientific support from the Institute of Medicine, now the Health and Medicine Division of the National Academies of Sciences. In 2004 it said:

> Initially the goal of infant formulas was to match the growth rate of the breastfed infant. However over time it was recognized that breastfeeding may confer several other potential advantages to the infant . . . including:
> - prevention of infectious diseases . . .,
> - neurodevelopment, and
> - protection from chronic diseases in childhood . . .
>
> These perceived and potential advantages of breast-feeding are the impetus behind many of the proposed additions of ingredients to infant formulas. Not all of these advantages are necessarily attributable to the nu-tritional content of human milk. Advantages resulting from a fundamentally different interaction between the nursing mother and her infant or to a selection bias of mothers who choose to breastfeed cannot be matched by simply adding nutrients to cow milk. (Institute of Medicine 2004, 7)

Evaluating infant formula mainly on whether it contains a good list of ingredients and leads to proper physical growth (height, weight) is far too simple. Unfortunately, no government agency anywhere has been clear about the various functions of infant feeding. All of them are important considerations in assessing the quality of infant feeding arrangements.

In previous publications I argued that infant formula must be not only safe but also *nutritionally adequate* (Kent 2011, 2012a; 2014b). I said we know formula is inadequate because, at the population level, infants who are fed with formula consistently have worse health outcomes than those who are breastfed. I now wish I had said these consistently worse health outcomes show the *overall* inadequacy of formula, and not just its nutritional inadequacy.

I was prompted to rethink my terminology when I reviewed a Power Point presentation by Stephen Buescher. I was stopped by the slide reproduced in Figure 8-1:

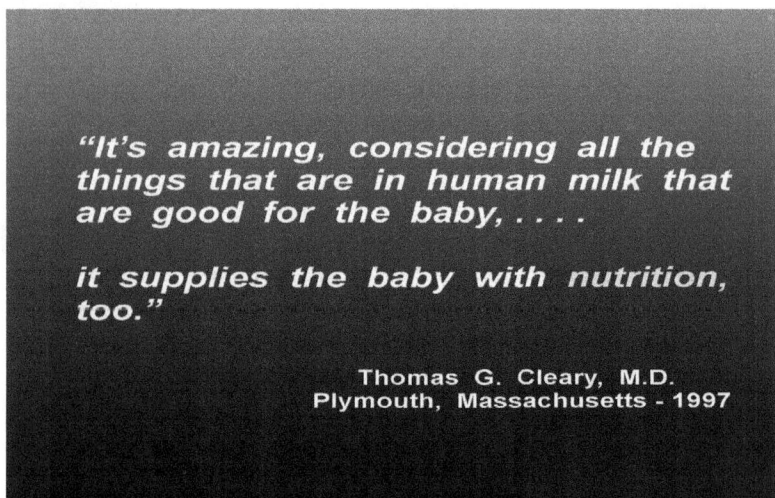

"It's amazing, considering all the things that are in human milk that are good for the baby, ...

it supplies the baby with nutrition, too."

Thomas G. Cleary, M.D.
Plymouth, Massachusetts - 1997

Figure 8-1. Human Milk Does Many Good Things,
and Also Provides Nutrition

The meaning is evident in a later slide in Buescher's presentation:

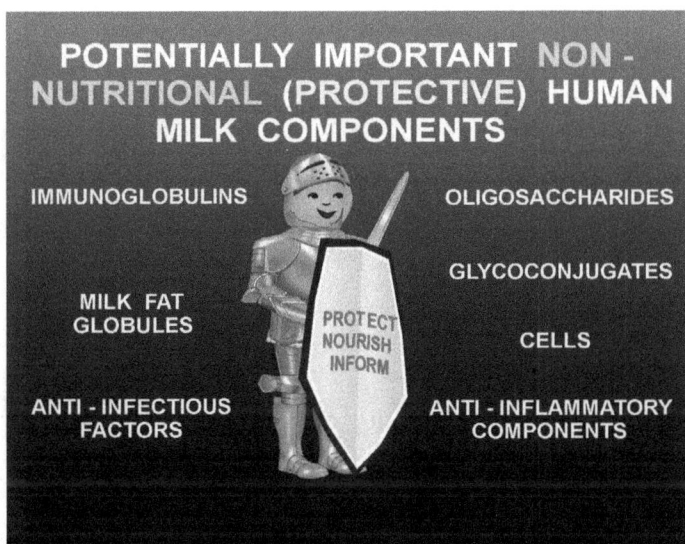

Figure 8-2. The Protective Components in Breast-milk

In this perspective, infant formulas could be nutritionally adequate while at the same time they are inadequate in other ways. As Buescher's analysis states, there are three main categories that need attention. Breastmilk provides not only *nourishment* but also *protective* and *informative* elements.

Figure 8-2 above lists the major protective elements. Some of the informative (communicative, signaling) functions are illustrated in Figure 8-3 below.

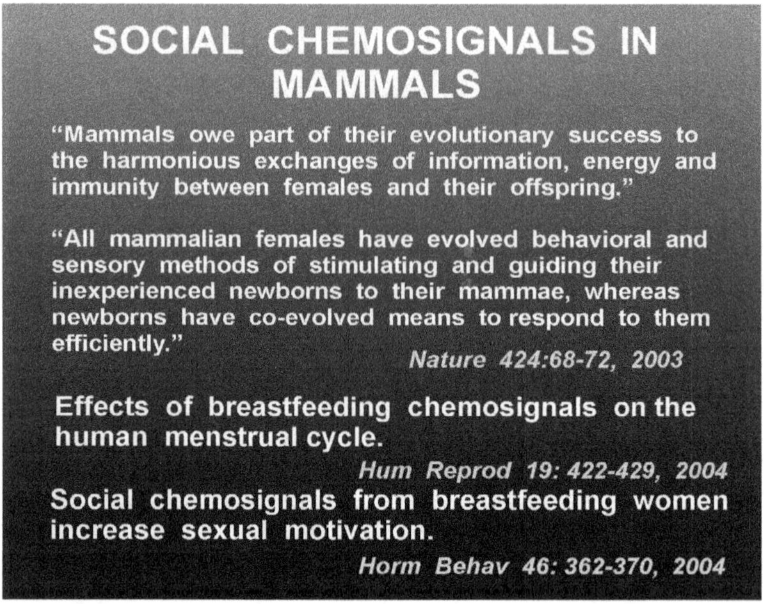

Figure 8-3. Signaling Between Mother and Infant

With direct breastfeeding, signals come from the infant to the mother about the infant's needs, and the mother can respond with suitably modified breastmilk. She might not be aware of that process, but her biology responds appropriately.

One mother voiced her awe about how this works, and accompanied her comments with the picture on the back cover of this book, to show that the milk she produces responds to her infant's signals (Lutkin 2016).

Here is how "breastmilk is custom-made to suit each baby":

> It varies nutritionally from country to country. A thirsty baby in hot weather will find more thirst-quenching feed at his mother's breast, if his mother drinks more

liquid, than a baby in a wintry climate. It varies with the time of day and even during the course of the feed: a baby's equivalent of soup to a rich dessert is available at one breast, while a drink to wash it down is waiting at the other. If the baby is extra hungry, and sucks more vigorously, the breast will obligingly produce larger helpings. A dainty eater's delicate sucking will inform the breast to dish up less. (Vittachi 1982)

A group of scientists offered this broad account of the multi-dimensional exchanges during breastfeeding:

> Mammals owe part of their evolutionary success to the harmonious exchanges of information, energy and immunity between females and their offspring. This functional reciprocity is vital for the survival and normal development of infants, and for the inclusive fitness of parents. It is best seen in the intense exchanges taking place around the mother's offering of, and the infant's quest for, milk. All mammalian females have evolved behavioural and sensory methods of stimulating and guiding their inexperienced newborns to their mammae, whereas newborns have coevolved means to respond to them efficiently. Among these cues, maternal odours have repeatedly been shown to be involved-but the chemical identity and pheromonal nature of these cues have not been definitively characterized until now. (Schaal et al. 2003; also see Blass 2003; Doucet et al. 2009)

According to another researcher's summary:

> Close body contact of the infant with his/her mother helps regulate the newborn's temperature, energy conservation, acid–base balance, adjustment of respiration, crying, and nursing behaviors. Similarly, the baby may

135

regulate—i.e., increase—the mother's attention to his/
her needs, the initiation and maintenance of breastfeeding, and the efficiency of her energy economy through
vagus activation and a surge of gastrointestinal tract
hormone release resulting in better exploitation of ingested calories. The effects of some of these changes can
be detected months later. (Winberg 2005)

There is a good illustration of the communication between infants
and mothers in the phenomenon of the "chest crawl" or "breast
crawl:

Every newborn, when placed on the mother's abdo-
men, soon after birth, has the ability to find its mother's
breast all on its own and to decide when to take the first
breastfeed. (Breast Crawl 2016)

There is no comparable formula crawl.

Many mothers sell their excess infant formula on the Internet. One
on Craig's List explained, "I have purchased various Enfamil infant
formula using coupons and my baby prefers breastmilk." Surely
that is evidence of clear communications from infant to mother. It
raises the question, why aren't infants' preferences considered when
policymakers make their policies?

Pediatricians and government regulators of infant formula often
focus on what they call nutrient adequacy. It is said to be achieved
if the list of ingredients in the formula match up with the list of
ingredients prescribed by an agency such as the Codex Alimentarius
Commission or the United States Food and Drug Administration.
Given its persistence, this approach can hardly be viewed as
accidental, a mistake. Assessing the quality of an infant's feeding
arrangements almost exclusively in terms of a list of ingredients
ingested must be viewed as a deliberate strategic choice. Officials
and experts who do that should explain how they justify it.

In the United States, the Institute of Medicine has played a particularly large role in supporting WIC's focus on specific nutrient components, not only in formula but also in WIC's food packages (Institute of Medicine 2005). Whether in examination of infant diets comprised solely of infant formula or complex diets for adults, the nutrient-based analyses should be just one part of much broader analyses of the quality of the diet.

Some experts do analyze infant feeding in terms of health outcomes. For example, a World Health Organization study assessed exclusive breastfeeding not in term of breastmilk's ingredients, but in terms of functional outcomes such as growth, immune response, and neurodevelopment (Butte, Lopez-Alarcon, and Garza, 2002). The researchers called it an assessment of nutrient adequacy, but many experts would say immune response is not about specific nutrients, but about dynamic biological processes (Goldman 2007).

If the experts themselves have such differing views, it is not surprising that others are easily confused.

Direct breastfeeding is a part of an active biological process involving dynamic interaction between mother and child. It is not a passive one-way process like filling a car's gas tank, where the only feedback is that loud click when the nozzle says the gas tank is full, thank you. There is a vast difference between infant formula and breastmilk, and another vast difference between delivering breastmilk through a bottle or tube and breastfeeding with direct warm contact between mother and child, each actively communicating with the other throughout the entire process. This dynamism is missing from the Institute of Medicine's way of viewing the infant feeding process.

We tend to think of eyes, ears, and skin as sensory organs. Breasts usually are not categorized that way, except in relation to sexuality.

In the context of infant feeding, often breasts are seen simply as passive fuel pumps, not as active instruments of communication.

The point here is comparable to Rudolf Steiner's argument that the human heart should not be viewed simply as a pump:

> The circulation of the blood is primary. Through its rhythmic pulsations—its systole and diastole—the heart responds to what takes place in the circulation of the blood. It is the blood that drives the heart and not the other way around. (Steiner, as quoted in Banner 2015, 267)

Nutrition should be understood as a process, not the contents of a package. To think of infant feeding as nothing more than a fuel transfer of a mix of unconnected ingredients trivializes everything and everyone involved. It is a bit like the notion that filling people's stomachs with cheap grains until they no longer feel hungry would somehow ensure they are well nourished.

There is a difference between saying "formula is worse than breastmilk," and saying "feeding with formula is worse than breastfeeding" for infants' and mothers' well-being. The first statement indicates a comparison between two different *products*. The second is about *processes*, the dynamic interaction between two highly active beings (Tonse 2011).

The breastfeeding process is embedded in an environment of complex processes. It is a bit like the many processes that go on simultaneously in a well-functioning car. No one shopping for a used car would simply examine the parts. The functions should be tested as well, including how well the car interacts with its driver. In breastfeeding, some processes are wholly within one human, and some involve interactions between two humans. Meaningful comparisons among different approaches to infant feeding should go beyond comparisons of ingredients and discuss the processes they are supposed to carry out. Does the feeding method being considered do what needs to be done?

Information about the way breastfeeding modifies breastmilk to meet infants' current needs is available to experts in academic journals, but they are not translated into non-scientific language and made readily available. One example of such a translation is that done by Hassiotou and her colleagues of their highly technical paper (Hassiotou et al. 2013) into a more reader-friendly one (Kakulas 2013; also see Bode et al. 2014; Newmark 2013).

Officials at high levels who support government subsidies of formula for infants or fail to ensure that women are fully supported in breastfeeding should know about formula's inadequacy. Their expert advisers should know. Not talking with new parents about these important dynamic processes in breastfeeding and connecting them with health consequences amounts to a conspiracy of silence.

Globally, infant formula has become such a big business that it has prompted the creation of a new sub-industry centered on the testing of infant formula (medGadget 2016). A 157-page overview of the industry is available for US$2850 (9Dimen Research 2016). Searching the Internet on "testing infant formula" brings to light many initiatives in formula testing. All focus on ingredients and economics, not formula's functions.

This sub-industry produces information for people in the formula business. It has little to say about infant health outcomes, or about comparisons between feeding with infant formula and breastfeeding. Who will fund and produce the objective and comprehensive information that new parents need when deciding how to feed their infants?

Recognizing their responsibilities, governments should take the lead in helping everyone, especially parents and policymakers, understand all the advantages and disadvantages of different methods of feeding infants. This expanded understanding should take account not only of health-related functions of the sort highlighted in this chapter (neurodevelopment, physical growth, resistance to infections, etc.), but also other important considerations such as health and other impacts on the mother, economic impacts on the family, religious and cultural considerations, convenience, and everything else that deserves fair consideration when making decisions about how to feed infants.

To end this book, I return to what was said in Chapter One:

In 1976, at its 11th session, the Codex Alimentarius Commission issued a *Statement on Infant Feeding*. It said, "it is necessary to encourage breastfeeding by all possible means in order to prevent that the decline in breastfeeding, which seems to be actually occurring, does not lead to artificial methods of infant feeding which could be inadequate or could have an adverse effect on the health of the infant (Codex Alimentarius Commission 1976)."

Forty years later, that summarizes the message of this book. The difference is that back then they thought artificial methods of infant feeding *could* have an adverse effect on the health of infants. Now we know that it does, with certainty. No government should subsidize it.

REFERENCES

9Dimen Research. 2016. *Global Infant Formula Testing Industry 2016, Trends and Forecast Report.* Market Research Reports.biz. http://www.marketresearchreports.biz/analysis/889684

Aguayo, Victor M., and Jay Ross. 2002. "The Monetary Value of Human Milk in Francophone West Africa: A PROFILES Analysis for Nutrition Policy Communication." *Food and Nutrition Bulletin*, 23 (2): 153-161. https://www.researchgate.net/publication/11280558_The_monetary_value_of_human_milk_in_Francophone_west_Africa_a_PROFILES_analysis_for_nutrition_policy_communication

Ahram Online. 2016. *Egyptians Protest in Cairo Over Shortages in Subsidised Baby Formula.* September 1. http://english.ahram.org.eg/NewsContent/1/64/242159/Egypt/Politics-/Egyptians-protest-in-Cairo-over-shortages-in-subsi.aspx

Akré, James E. 2012. "Human-milk Sharing: A Timeless Practice Goes Mainstream." *Global Online Lactation Discussion*, Gold. May 4. 12. https://www.google.com/url?sa=t&rct=j&q=&esrc=s&source=web&cd=1&ved=0ahUKEwjshfbmoKbQAhXprFQKIIW3iAbsQFggdMAA&url=https%3A%2F%2Fwww.researchgate.net%2Ffile.PostFileLoader.html%3Fid%3D5579bc635e9d97dc8d8b456a%26assetKey%3DAS%253A273794432733188%25401442289087858&usg=AFQjCNFo8iUC7U__OfSAPAXmjCgMcoTdqA

Akré, James E., Karleen D. Gribble, and Maureen Minchin. 2011. "Milk sharing: from private practice to public pursuit." *International Breastfeeding Journal*, 6 (8). http://www.internationalbreastfeedingjournal.com/content/6/1/8

Al Ghad, Amwal. 2016. "Doctors Syndicate Call for Lowering Subsidised Milk Price to EGP5." *Egypt News*. September 8. http://www.amwalalghad.com/en/?tmpl=component&option=com_content&id=50390

Alipui, Nicholas 2011. *UNICEF expresses its support for the Resolution opposing the health claim.* Babymilk Action. http://info.babymilkaction.org/news/policyblog/unicefDHA

Al-Naqeeb, Niran, Ayman Azab, Mahmoud S. Eliwa, and Bothaina Y. Mohammed. 2000. "The Introduction of Breast Milk Donation in a Muslim Country." *Journal of Human Lactation* December. 16 (4): 346-50. https://www.researchgate.net/publication/12173231_The_Introduction_of_Breast_Milk_Donation_in_a_Muslim_Country

Alonso-Zaldivar. 2016. "Did Landmark Laws from Congress Enable High Drug Prices." *AP: The Big Story*. September 29. http://bigstory.ap.org/article/e0a63c0803164ac59f19cb1fa1d8c5c7/did-landmark-laws-congress-enable-high-drug-prices

American Pregnancy Association. 2016. *What's in Breast Milk?* http://americanpregnancy.org/first-year-of-life/whats-in-breastmilk/

Ambrosia Labs. 2016. Website. http://ambrosiamilk.com/

AMEInfo. 2015. *Saudi Arabia pays SAR8 million to subsidize formula, fodder, and barley.* AMEinfo. August 18. http://ameinfo.com/money/economy/saudi-arabia-pays-sar8-billion-to-subsidise-formula-fodder-and-barley/

American Academy of Pediatrics. 2012. "Breastfeeding and Human Milk." *Pediatrics.* March. 129 (3). http://pediatrics.aappublications.org/content/129/3/e827.full#content-block

American Academy of Pediatrics. Committee on Pediatric AIDS. 2013. "Infant Feeding and Transmission of Human Immunodeficiency Virus in the United States." *Pediatrics.* 131 (2). http://pediatrics.aappublications.org/content/131/2/391

Arnold, Lois D. W. 2006. "Global health policies that support the use of banked donor human milk: a human rights issue." *International Breastfeeding Journal.* December 12. 1 (26). http://www.internationalbreastfeedingjournal.com/content/pdf/1746-4358-1-26.pdf

Astley, Mark. 2014. "Saudi Arabian authorities have warned seven domestic and international infant formula manufacturers and importers that they face severe penalties for illegally driving up the price of their products." *DairyReporter.com* http://mobile.dairyreporter.com/Manufacturers/Infant-formula-firms-facing-fines-for-Saudi-Arabian-price-hikes

Aswat Masriya. 2016a. *Egypt's Armed Forces to Supply Baby Formula at Reduced Prices—Army Spokesman.* All Africa. September 3. http://allafrica.com/stories/201609040039.html

---. 2016b. "'New System' of Infant Formula Sales Introduced in Egypt Amid Protest Over Shortages." *Egyptian Streets.* September 2. http://egyptianstreets.com/2016/09/02/new-system-of-infant-formula-sales-introduced-in-egypt-amid-protests-over-shortages/

Aunpalmquist 2014. "Taking milk from strangers." *Anthrolactology.* December 29. https://anthrolactology.wordpress.com/author/aunpalmquist/

Badger, Thomas M., Janet M. Gilchrist, R. Terry Pivik, Aline Andres, Kartik Shankar, Jin-Ran Chen. and Martin J. Ronis. 2009. "The health implications of soy infant formula." *American Journal of Clinical Nutrition.* 89 (5 suppl):1668S-1672S. http://www.ajcn.org/content/89/5/1668S.full

Baker, Phillip Ian, Julie Patricia Smith, Libby Salmon, Sharon Friel, George Kent, Alessandro Iellemo, Jai Prakash Dadhich, and Mary J. Renfrew. 2016. "Global trends and patterns of commercial milk-based formula sales: is an unprecedented infant and young child feeding transition underway?" *Public Health Nutrition.* May. http://papers.ssrn.com/sol3/papers.cfm?abstract_id=2786419

Barber, Dan. 2016. *The Third Plate: Field Notes on the Future of Food.* New York: Penguin Books.

Barruti, Soledad. 2016a. "Bad Milk: Breastfeeding, Baby Formula, and Business Interests. *The Argentina Independent.* November 22. http://www.argentinaindependent.com/currentaffairs/analysis/bad-milk-breastfeeding-baby-formula-and-business-interests/

Barruti, Soledad. 2016b. "Mala Leche: Otro Negocio que Entrega el Estado a las Corporaciones." *Mu: El :Periódico de La Vaca.* November. 10 (105). http://www.lavaca.org/notas/mala-leche-otro-negocio-que-entrega-el-estado-a-las-corporaciones/

Bartick, Melissa C., and Arnold Reinhold. 2010. "The Burden of Suboptimal Breastfeeding in the United States: A Pediatric Cost Analysis." *Pediatrics.* 125(5): e1048-56. http://www.breastfeedingor.org/wp-content/uploads/2012/10/burden-of-suboptimal-breastfeeding-in-the-us.-a-cost-analysis.pdf

Bartick, Melissa, C., Eleanor Bimla Schwarz, Brittany D. Green, Briana J. Jegier, Arnold G. Reinhold, Tarah T. Colaizy, Debra L. Bogen, Andrew J. Schaefer, and Alison M. Stuebe. 2016. "Suboptimal Breastfeeding in the United States: Maternal and Pediatric Health Outcomes and Costs." *Maternal & Child Nutrition.* September 19. http://onlinelibrary.wiley.com/doi/10.1111/mcn.12366/full

Baumgartel, Kelley L., and Diane L. Spatz. 2013. "WIC (The Special Supplemental Program for Women, Infants, and Children): Policy Versus Practice Regarding Breastfeeding." *Nursing Outlook.* 61(6): 466-70. http://www.nursingoutlook.org/article/S0029-6554(13)00141-3/fulltext

Baumslag, Naomi, and Dia L. Michels. 1995. *Milk, Money, and Madness: The Culture and Politics of Breastfeeding.* Westport, Connecticut: Bergin & Garvey.

Belamarich, Peter F., Risa E. Bochner, and Andrew D. Racine
. 2015. "A Critical Review of the Marketing Claims of
Infant Formula Products in the United States." *Clinical
Pediatrics*. May; 55 (5):437-42. https://blogs.deakin.
edu.au/apfnc/wp-content/uploads/sites/119/2015/06/
Belamarich_2015_A-Critical-Review-of-the-Marketing-
Claims-of-Infant-Formula-Products-in-the-United-
States.-Clin-Pediatr.pdf

Belluck, Pam. 2015. "Premature Babies May Survive at 22 Weeks
if Treated, Study Finds." *New York Times*, May 6. http://
www.nytimes.com/2015/05/07/health/premature-babies-
22-weeks-viability-study.html?emc=edit_th_20150507&
nl=todaysheadlines&nlid=2155033&_r=0

Benefit Corporation. 2016. Website. http://benefitcorp.net/

Berg, Alan. 1973. *The Nutrition Factor*. Washington, D.C.:
Brookings Institute.

Besharov, Douglas J. and Germanis, Peter. 2000. "Evaluating
WIC," *Evaluation Review*. 24 (2)2 123-190. http://erx.
sagepub.com/content/24/2/123.abstract

Black, Andrew P., Julie Brimblecombe, Helen Eyles, Peter Morris,
Hassan Vally, and Kerin O'Dea. 2012. "Food Subsidy
Programs and the Health and Nutritional Status of
Disadvantaged Families in High Income Countries: A
Systematic Review." *BMC Public Health*. 12: 1099. doi:
10.1186/1471-2458-12-1099 or http://europepmc.org/
articles/PMC3559269

Black, Maureen M., Diana B. Cutts, Deborah A. Frank, Joni Geppert, Anne Skalicky, Suzette Levenson, Patrick H. Casey, Carol Berkowitz, Nieves Zaldivar, John T. Cook, Alan F. Meyers, and Tim Herren. 2004. "Special Supplemental Nutrition Program for Women, Infants, and Children Participation and Infants' Growth and Health: A Multisite Surveillance Study." Pediatrics. July. 114 (1). http://pediatrics.aappublications.org/content/114/1/169

Blass, Elliott M. 2003. "Reproductive Biology: Mammary Messages." *Nature*. 424 (25-26).

Boban, Marajo and Irena Zakarija-Grković. 2016. "In-Hospital Formula Supplementation of Healthy Newborns: Practices, Reasons, and Their Medical Justification." *Breastfeeding Medicine*. August 22. http://online.liebertpub.com/doi/pdfplus/10.1089/bfm.2016.0039

Bode, Lars, Mark McGuire, Juan M. Rodriguez, Donna T. Geddes, Foteini Hassiotou, Peter E. Hartmann, and Michelle K. McGuire. 2014. "It's Alive: Microbes and Cells in Human Milk and Their Potential Benefits to Mother and Infant." *Advances in Nutrition*. oi:10.3945/an.114.006643. http://m.advances.nutrition.org/content/5/5/571.full.pdf

Bologna, Caroline. 2016. "Mom Schools the Internet on the Magic of Breast Milk in Viral Post." *Huffington Post*. February 23. http://www.huffingtonpost.com.au/entry/mom-schools-the-internet-on-the-magic-of-breast-milk-in-viral-post_us_56cb8c8fe4b041136f180759?ncid=fcbklnkushpmg00000046&section=australia

Brady, June Pauline 2012. "Marketing Breast milk Substitutes: Problems and Perils Throughout the World." *Archives of Disease in Childhood.* 97: 529-532. http://adc.bmj.com/content/97/6/529.full.html

Breast Crawl. 2016. Website. http://www.breastcrawl.org/

Buia, Carolina. 2015. "The Booming Market for Breast Milk." *Newsweek.* May 23. http://www.newsweek.com/booming-market-breast-milk-335151

Butte, Nancy F., Mardia G. Lopez-Alarcon, and Cutberto Garz. 2002. *Nutrient Adequacy of Exclusive Breastfeeding for the Term Infant During the First Six Months of Life.* Geneva: World Health Organization. http://www.who.int/nutrition/publications/infantfeeding/9241562110/en/

Bye, Clarissa. 2016. "Mums Desperate for Breast Milk Turn to Internet Due to Shortage of Recommended Milk Banks." *Daily Telegraph.* July 29. http://www.dailytelegraph.com.au/news/nsw/mums-desperate-for-breast-milk-turn-to-internet-due-to-shortage-of-recommended-milk-banks/news-story/06fa7eb53468fd361c84410382c43d2e

California WIC Association. 2010. *Concerns About Infant Formula Marketing and Additives.* http://www.calwic.org/storage/documents/federal/2010/formulabrief.pdf

Campbell, Olivia. 2016. "When Babies Need Donated Breast Milk, Should States Pay?" *Scientific American: Stat.* October 4. https://www.statnews.com/2016/10/04/breast-milk-laws-premature-babies/

Carlson, Steven, Robert Greenstein, and Zoë Neuberger. 2015. *WIC's Competitive Bidding Process for Infant Formula is Highly Cost Effective.* September 14. Washington, D.C.: Center on Budget and Policy Priorities. http://www.cbpp. org/research/food-assistance/wics-competitive-bidding-process-for-infant-formula-is-highly-cost

Carlson, Steven and Zoë Neuberger. 2015. *WIC Works: Addressing the Nutrition and Health Needs of Low-Income Families for 40 Years.* May 4. Washington, D.C.: Center on Budget and Policy Priorities.

Carter, Shannon K., Beatriz Reyes-Foster, and Tiffany L. Rogers 2015. "Liquid Gold or Russian Roulette? Risk and Human Milk Sharing in the US News Media." *Health, Risk & Society.* Vol. 17, No 1, 30-45. http://www. tandfonline.com/doi/abs/10.1080/13698575.2014.1000 269

Castillo, Cecilia, and Leslie Power. 2016. "¿Cuida el Ministerio de Salud de Chile la Lactancia Materna?" *Opinión Publicado.* Centro de Investigación Periodística. July 14. http:// ciperchile.cl/2016/07/14/cuida-el-ministerio-de-salud-de-chile-la-lactancia-materna/

Center on Budget and Policy Priorities. 2015. *Policy Basics: Special Supplemental Nutrition Program for Women, Infants, and Children.* February 9. Washington, D.C.: CBPP. http://www.cbpp.org/research/policy-basics-special-supplemental-nutrition-program-for-women-infants-and-children

CEPPs. 2016. *Breastfeeding*. /Childhood and Early Parenting Principles. http://cepprinciples.org/breastfeeding/

Chen, Aimin, and Walter J. Rogan. 2004. Breastfeeding and the Risk of Postneonatal Death in the United States. *Pediatrics* 113:5 (May) 435-439. http://pediatrics. aappublications.org/content/113/5/e435

Coburn, Jennifer. 2000. "Formula for Profit: How Marketing Breastmilk Substitutes Undermines the Health of Babies." *Mothering*. July/August. https://www.researchgate.net/ publication/265536691_Formula_for_Profit_How_ Marketing_Breastmilk_Substitutes_Undermines_the_ Health_of_Babies

Codex Alimentarius Commission. 1976. *Statement on Infant Feeding, CAC/MISC-2-1976*. http://www. codexalimentarius.net/download/standards/301/ CXA_002e.pdf

---. 1987. Codex Standard for Follow-Up Formula: CodeX Stan 156-1987. http://www.fao.org/fao-who-codexalimentarius/sh-proxy/en/?lnk=1&url=https%253A %252F%252Fworkspace.fao.org%252Fsites%252Fcode x%252FStandards%252FCODEX%2BSTAN%2B156-1987%252FCXS_156e.pdf

---. 1989. *Codex General Guidelines for the Utilization of Vegetable Protein Products (VPP) in Foods*. CAC/GL 4-1989. http:// www.fao.org/fao-who-codexalimentarius/sh-proxy/ru/?ln k=1&url=https%253A%252F%252Fworkspace.fao.org% 252Fsites%252Fcodex%252FStandards%252FCAC%2B GL%2B4-1989%252FCXG_004e.pdf

---. 2007. *Standards for Infant Formulas and Formulas for Special Medical Purposes Intended for Infants (Codex Stan 72-1981) (Revised 2007)*. http://npaf.ca/wp-content/uploads/2014/02/CXS_072E-Codex-Standard-Infant-Formula-.pdf

---. 2016. Website. http://www.fao.org/fao-who-codexalimentarius/en/

Commonwealth of Australia. 2014. *Donor Human Milk Banking in Australia—Issues and Background Paper*. Canberra, Australia: Commonwealth of Australia. https://www.google.com/url?sa=t&rct=j&q=&esrc=s&source=web&cd=1&ved=0CCgQFjAA&url=http%3A%2F%2Fwww.health.gov.au%2Finternet%2Fmain%2FPublishing.nsf%2FContent%2FD94D40B034E00B29CA257BF0001C AB31%2F%24File%2FDonor%2520Human%2520Mil k%2520Banking%2520in%2520Australia%2520paper% 2520%2528D14-1113484%2529.docx&ei=49o1VYO6 PI7joASPhYCIBg&usg=AFQjCNGILP3MSj31hu9cnfX-jjYyCVEr6A&bvm=bv.91386359,d.cGU

Davis, David E., and Victor Oliveira. 2015. *Manufacturers' Bids for WIC Infant Formula Rebate Contracts, 2003-2013*. Washington, D.C.: United States Department of Agriculture. July. Economic Information Bulletin. Number 142. http://www.ers.usda.gov/media/1863322/eib142.pdf

Ginty, Molly M. 2011. *Infant-Formula Companies Milk U.S. Food Program*. *WomensEnews*. November 7. http://womensenews.org/2011/11/infant-formula-companies-milk-us-food-program/

Dawson, David Stephanie. 2011. "Legal Commentary on the Internet Sale of Human Milk." *Public Health Reports.* March-April.126 (1). http://www.ncbi.nlm.nih.gov/pmc/articles/PMC3056027/

de Wagt, Arjan, and David Clark. 2014. *UNICEF's Support to Free Infant Formula for Infants of HIV-Infected Mothers in Africa: A Review of UNICEF Experience.* Penang, Malaysia: World Alliance for Breastfeeding Action. http://www.waba.org.my/whatwedo/hcp/pdf/r-unicef-free.pdf

Digger History. 2016. *Coca-Cola at War (on BOTH sides).* http://www.diggerhistory.info/pages-food/coca_cola.htm

Doucet, Sébastian, Robert Soussignan, Paul Sagot, and Benoist Schaal. 2009. "The Secretion of Areolar (Montgomery's) Glands from Lactating Women Elicits Selective, Unconditional Responses in Neonates." *PLOS One.* October 23. http://dx.doi.org/10.1371/journal.pone.0007579

Drago, Robert. 2011. "The WIC Program: An Economic Analysis of Breastfeeding and Infant Formula." *Breastfeeding Medicine.* 6 (5): 281-286.

Dutton, Judy 2011. "Liquid Gold: The Booming Market for Human Breast Milk." *Wired,* May 17. http://www.wired.com/2011/05/ff_milk/all/1

Eats on Feets 2015. Website. www.eatsonfeets.org

Ecker, Olivier, Jean Francois Trinh Tan, and Perrihan Al-Riffai. 2014. *Facing the Challenge: The Recent Reform of the Egyptian Food Subsidy System.* Arab Spatial. December 19. http://www.arabspatial.org/blog/blog/2014/12/19/facing-the-challenge-the-recent-reform-of-the-egyptian-food-subsidy-system/

Ecker, Olivier, Perrihan Al-Riffai, Clemens Breisinger, and Rawia El-Bagrawy. 2016. *Nutrition and Economic Development: Exploring Egypt's Exceptionalism and the Role of Food Subsidies.* Washington, D.C.: International Food Policy Research Institute. http://www.ifpri.org/publication/ nutrition-and-economic-development-exploring-egypts- exceptionalism-and-role-food-0

EDHS. 2014. *Egypt Demographic and Health Survey 2014: Main Findings.* Ministry of Health and Population, El-Zanaty and Associates, The DHS Program. http://dhsprogram. com/pubs/pdf/PR54/PR54.pdf

Edmond, Karen M., Betty R. Kirkwood, Amenga-Etego Seeba, Seth Owusu-Agyei, Lisa S. Hurt. 2007. "Effect of Early Infant Feeding Practices on Infection-specific Neonatal mortality: An Investigation of the Causal Links with Observational Data from Rural Ghana." *American Journal of Clinical Nutrition.* October. 864, 1126-1131. http:// ajcn.nutrition.org/content/86/4/1126.long

Egyptian Pharmaceutical Trading Company. 2016. Website. http://www.eptc-egydrug.com/arb_Page48.htm

El-Bar, Karim. 2016. "Got milk? Egypt to Check Women's Breasts Before Supplying Baby Formula." *Middle East Eye.* http://www.middleeasteye.net/news/got-milk- egypt-wants-check-women-s-breasts-supplying-baby- formula-1819401048

El-Tablawy, Tarek. 2016. "When the Baby Milk Disappeared, Egypt Turned to the Military." *Bloomberg.* September 8. http://www.bloomberg.com/news/articles/2016-09-08/ when-the-baby-milk-disappeared-egypt-turned-to-the- military

El-Zanaty, Fatma, and Ann Way. 2001. *Egypt Demographic and Health Survey 2000*. http://dhsprogram.com/pubs/pdf/ FR117/00FrontMatter.pdf

---. 2009. *Egypt Democratic and Health Survey 2008*. https:// dhsprogram.com/pubs/pdf/FR220/FR220.pdf

EMBA 2011. *The Sharing of Human Milk*. European Milk Bank Association. http://www.europeanmilkbanking.com/ images/news/Human milk%20Sharing.%20A%20 statement%20from%20the%20European%20Milk%20 Bank%20Association%20December%202011.pdf

ENN. 2016. *Emergency Nutrition Network*. Website. http://www. ennonline.net

Ettinger, Jill. 2016. "Healthy Eating Habits Created Through Government Assistance Program." *Organic Authority*. October 7. http://www.organicauthority.com/healthy- eating-habits-created-through-government-assistance- program/

Farid, Doaa. 2016. "70 Million Reliant on Food Subsidies in Peril amid Government Austerity and Mismanagement." *Daily News: Egypt*. March 2. http://www.dailynewsegypt. com/2016/03/02/70-million-reliant-on-food-subsidies-in- peril-amid-government-austerity-and-mismanagement/

Fentiman, Linda C. 2010. "Marketing Mother's Milk: The Commodification of Breastfeeding and the New Markets for Breast Milk and Infant Formula." *Nevada Law Journal*. 10(1). http://scholars.law.unlv.edu/nlj/vol10/ iss1/3/

Formoso, Jessica. 2016. *Donor Breast Milk Not Always Covered by Health Insurance*. Fox 5. October 19. http://www.fox5ny. com/news/212474061-story

Forsyth, Steward. 2013. "Non-compliance with the International Code of Marketing of Breast Milk Substitutes is not Confined to the Infant Formula Industry." *Journal of Public Health*. 35(2): 185-190. http://jpubhealth. oxfordjournals.org/content/early/2013/05/07/pubmed. fds084.full

Francis, Solveig, Selma James, Phoebe Jones Schellenberg, and Nina Lopez-Jones 2002. *The Milk of Human Kindness*. London: Crossroads.

Ghaly, Mohammed. 2010. "Milk Banks Through the Lens of Muslim Scholars: One Text in Two Contexts." *Bioethics*. doi:10.1111/j.1467-8519.2010.01844.x. https:// openaccess.leidenuniv.nl/bitstream/handle/1887/16542/ Ghaly-Milk%20Banks.pdf?sequence=1

Ghosh, Mrinal, Virginia Nguyen, H. Konrad Muller, and Amae M. Walker. 2016. "Maternal Milk T Cells Drive Development of Transgenerational Th1 Immunity in Offspring Thymus." *Journal of Immunology*, 2016; 197 (6): 2290 DOI: 10.4049/%u200Bjimmunol.1502483 http://www.jimmunol.org/content/early/2016/08/04/ jimmunol.1502483.abstract

Ginty, Molly M. 2011. *Infant-Formula Companies Milk U.S. Food Program*. Women's eNews. November 7. http:// womensenews.org/2011/11/infant-formula-companies-milk-us-food-program/

Golden, Janet 2001. *A Social History of Wet Nursing in America: From Breast to Bottle*. Cambridge: Cambridge University Press.

Goldman, Armond S. 2007. "The Immune System in Human Milk and the Developing Infant." *Breastfeeding Medicine.* 2 (4): 195-204. https://www.ncbi.nlm.nih.gov/pubmed/18081456

GoodGuide. 2011. *Six Secrets About Infant Formula.* August 23. https://blog.goodguide.com/2011/08/23/six-secrets-about-infant-formula/

Greiner, Ted. 2011. "The Free Lunch is Always an Effective Marketing Tool: Why WIC Must Change." Letter to the Editor. *Breastfeeding Medicine.* 6. http: DOI: 10.1089/bfm.2011.0050

Gribble, Karleen D. and Bernice L. Hausman. 2012. "Milk Sharing and Formula Feeding: Infant Feeding Risks in Comparative Perspective." *Australasian Medical Journal.* 5 (5): 275-283. http://dx.doi.org/10.4066/AMJ.2012.1222

Grummer-Strawn, Laurence M. and Nigel Rollins. 2015. "Summarising the Health Effects of Breastfeeding." Acta Paediatrica, December. DOI: 10.1111/apa.13136 http://onlinelibrary.wiley.com/doi/10.1111/apa.13136/epdf

Handley, Erin, and Kong Meta. 2016. "Formula Marketers Accused of Skating Laws That Encourage Breastfeeding." *Phnom Penh Post.* August 15. http://www.phnompenhpost.com/national/formula-marketers-accused-skating-laws-encourage-breastfeeding

Hassiotou, Foteini, Anna R. Hepworth, Philipp Metzger, Ching Tai Lai, Naomi Trengove, Peter E. Harmann, and Luis Figueira. 2013. "Maternal and Infant Infections Stimulate a Rapid Leukocyte Response in Breastmilk." *Clinical & Translational Immunology.* April. 2 (e3). http://www.nature.com/cti/journal/v2/n4/pdf/cti20131a.pdf

Hatløy, Anne, and Arne Oshaug 1997 "Human Milk: An Invisible Food Source. *Journal of Human Lactation.* December. 13 (4): 299-305.

Heise, Gudrun 2014. "Germany's New Online Breast Milk Marketplace." *DW.* http://www.dw.de/germanys-new-online-breast-milk-marketplace/a-17838956

hm4hb 2014. *Human Milk for Human Babies Global Network.* Website. http://www.hm4hb.net/

Holla, Radha, Alessandro Iellamo, Arun Gupta, Julie Smith, and J.P. Dadhich. 2013. *The Need to Invest in Babies.* International Baby Food Action Network and Breastfeeding Promotion Network of India. http://infactcanada.ca/pdf/the-need-to-invest-in-babies.pdf

Holmes, Alison Volpe, Nancy P. Chin, Jeffery Kaczorowski, and Cindy R. Howard. 2009. "A Barrier to Exclusive Breastfeeding for WIC Enrollees: Limited Use of Exclusive Breastfeeding Food Package for Mothers." *Breastfeeding Medicine.* March. 4 (1): 25-30.

IBFAN. 2011. *Egypt Code Violations 2011.* Look What They're Doing Penang, Malaysia: International Baby Food Action Network. http://ibfan.org/art/LWTD-Egypt.pdf

---. 2003. *Egypt Code Violations 2003.* Penang, Malaysia: International Baby Food Action Network. http://ibfan.org/art/298-3.pdf

---. 2016. International Baby Food Action Network. Website. http://www.ibfan.org/

ICDC 2016. International Code Documentation Centre. Website. http://www.ibfan-icdc.org/

IFPRI. 2013. *Tackling Egypt's Rising Food Insecurity in a Time of Transition*. International Food Policy Research Institute. May. http://ebrary.ifpri.org/utils/getfile/collection/ p15738coll2/id/127559/filename/127770.pdf

Infant Feeding Support for Refugee Children. 2016. *You Can Help Refugee Mothers Safely Feed Their Children*. January 29. https://www.facebook.com/notes/infant-feeding-support-for-refugee-children/you-can-help-refugee-mothers-safely-feed-their-children/212918912388107

Institute of Medicine. 1996. *WIC Nutrition Risk Criteria: A Scientific Assessment*. Washington, D.C.: National Academy Press.

---. 2004. "Comparing Infant Formulas with Human Milk. Chapter 3. *Infant Formula: Evaluating the Safety of New Ingredients. Washingto*n, D.C.: The National Academies Press. https://www.ncbi.nlm.nih.gov/books/NBK215837/

---. 2005. "Nutrient and Food Priorities for the WIC Food Packages." Chapter 2. *WIC Food Packages: Time for a Change*. Washington, D.C.: The National Academies Press. https://www.nap.edu/read/11280/chapter/4

Ip, Stanley, Mei Chung. Gowri Raman, Priscilla Chew, Nombulelo Magula, Deirdre DeVine, Thomas Trikalinos, Joseph Lau, 2007. *Breastfeeding and Maternal and Infant Health Outcomes in Developed Countries*. Rockville, Maryland: Agency for Healthcare Research and Quality, U.S. Department of Health and Human Services. https:// archive.ahrq.gov/downloads/pub/evidence/pdf/brfout/ brfout.pdf

JAMA. 1925. "Dried Human Milk." JAMA (Journal of the American Medical Association) 86 (5): 352-353. http:// jama.jamanetwork.com/article.aspx?articleid=239441

Johannes, Laura. 2015. "Should Breast Milk Be Nutritionally
     Analyzed?" *Wall Street Journal.* December 28. http://www.
     wsj.com/articles/should-breast-milk-be-nutritrionally-
     analyzed-1451319691

Joint Statement. 2016. *Joint Statement by the UN Special
     Rapporteurs on the Right to Food, Right to Health, the
     Working Group on Discrimination Against Women in Law
     and in Practice, and the Committee on the Rights of the
     Child in Support of Increased Efforts to Promote, Support
     and Protect Breast-feeding.* Geneva: Office of the High
     Commissioner for Human Rights. http://www.ohchr.org/
     EN/NewsEvents/Pages/DisplayNews.aspx?NewsID=2087
     1&LangID=E#sthash.QdT22h0B.dpuf

Journey Staff. 2012. *The Chronicle of Coca-Cola: A Symbol of
     Friendship.* Coca-Cola Journey. January 1. http://www.
     coca-colacompany.com/stories/the-chronicle-of-coca-cola-
     a-symbol-of-friendship

Kakulas (formerly Hassiotou), Foteini. 2013 "Protective Cells
     in Breast Milk: For the Infant and the Mother?" *Splash!*
     International Milk Genomics Consortium. April. http://
     milkgenomics.org/article/protective-cells-in-breast-milk-
     for-the-infant-and-the-mother/

Kaplan, Kimberly. 2009. "Salma Hayek on Why She
     Breastfed Another Woman's Baby." *ABC News.*
     February 11. http://abcnews.go.com/Entertainment/
     story?id=6854285&page=1

KellyMom. 2015. *Why Keep Infant Formula Marketing Out of
     Healthcare Facilities?* KellyMom Parenting/Breastfeeding.
     http://kellymom.com/blog-post/why-keep-infant-
     formula-marketing-out-of-healthcare-facilities/

Kent, George. 1995. *Children in the International Political Economy.* London/New York: Macmillan/St. Martin's.

---. 1997."Realizing Infants' Nutrition Rights Through Implementation of National Law," *International Journal of Children's Rights.* 5 (4): 439-456. http://www2.hawaii.edu/~kent/IJCR1997KentRealizingChildrensRights.pdf

---. 1998. "Women's Rights to Breastfeed vs. Infants' Rights to be Breastfed," *SCN News* (United Nations Sub-Committee on Nutrition). 17: 18-19. http://www.unsystem.org/scn/Publications/SCNNews/SCNNews17.pdf

---. 2001. "Breastfeeding: A Human Rights Issue?" *Development.* 44 (2) 93-98. http://www2.hawaii.edu/~kent/breastfeedingrights.pdf

---. 2004. "Human Rights and Infant Nutrition," *WABA Global Forum II-23-27 September 2002-Arusha, Tanzania.* Penang, Malaysia: World Alliance for Breastfeeding Action. pp. 178-186. http://www2.hawaii.edu/~kent/HUMAN%20RIGHTS%20AND%20INFANT%20NUTRITION.pdf

---. 2005a *Freedom from Want: The Human Right to Adequate Food.* Washington, D.C.: Georgetown University. http://press.georgetown.edu/sites/default/files/978-1-58901-055-0%20w%20CC%20license.pdf

---. 2005b. "HIV/AIDS, Infant Feeding, and Human Rights." In Wenche Barth Eide and Uwe Kracht, eds., *Food and Human Rights in Development. Volume I. Legal and Institutional Dimensions and Selected Topics* (Antwerp, Belgium: Intersentia.

---. 2006a. "Child Feeding and Human Rights." *International Breastfeeding Journal.* 1: 27. https://www.ncbi.nlm.nih.gov/pmc/articles/PMC1764724/

---. 2006b. "WIC's Promotion of Infant Formula in the United States." *International Breastfeeding Journal.* 1 (8). April. http://www.internationalbreastfeedingjournal.com/content/1/1/8

---. 2011. *Regulating Infant Formula.* Amarillo, Texas: Hale Publishing.

---. 2012a. "The Nutritional Adequacy of Infant Formula." *Clinical Lactation,* 3 (1): 21-25. http://www.ingentaconnect.com/content/springer/clac/2012/00000003/00000001/art00004

---. 2012b. "Ultra-Processed Products: The Trouble Starts with Baby Formula." *World Nutrition.* October. 3. (10): 449-455. http://www.wphna.org/htdocs/downloadsoct2012/12-10_WN3_breastfeeding_pdf.pdf

---. 2014a. "Regulating fatty acids in infant formula: critical assessment of U.S. policies and practices." *International Breastfeeding Journal.* 9 (2). http://www.ncbi.nlm.nih.gov/pmc/articles/PMC3922793/

---. 2014b. "Regulating the Nutritional Adequacy of Infant Formula in the United States." *Clinical Lactation,* Vol. 5, No. 4, pp. 133-136 (2014). http://www.ingentaconnect.com/content/springer/clac/2014/00000005/00000004/art00006

---. 2014c. *Quality Factors in New Infant Formula Requirements.* Comment on proposed rule of the US Food and Drug Administration. June 23, 2014.

---. 2015. "Global Infant Formula: Monitoring and Regulating the Impacts to Protect Human Health." *International Breastfeeding Journal.* 9.

---. 2016. *Caring About Hunger.* Sparsnäs, Sweden: Irene Publishing.

Kent, George, and Cecilia Castillo. 2016. "Government Distribution of Infant Formula in Chile." Social Medicine. 10 (3). September. In English: http://socialmedicine.info/index.php/socialmedicine/article/view/900/1690, pages 76-82. In Spanish: http://www.medicinasocial.info/index.php/medicinasocial/article/view/903, pages 82-89.

Kent, George, Ghada Sayed and Azza Abul-Fadl. 2017. "Government Provision of Infant Formula in Egypt." Social Medicine. 11 (1): January. In English: http://socialmedicine.info/index.php/socialmedicine/article/view/920/1732, pages 1-7. In Spanish: http://www.medicinasocial.info/index.php/medicinasocial/article/view/923/1724, pages 1-7.

Khalil, Aliaa, Rachel Buffin, Damien Sanlaville' and Jean-Charles Picaud. 2016. "Milk kinship is not an obstacle to using donor human milk to feed preterm infants in Muslim countries." *Acta Pediatrica.* May. 105 (5): 462-467. http://onlinelibrary.wiley.com/wol1/doi/10.1111/apa.13308/full

Labbok, Miriam. 2015. "Paying Moms for their Milk – A Slippery Slope." *Breastfeeding4Health.* January 12. http://www.breastfeeding4health.com/2015/01/paying-moms-for-their-milk-slippery.html

Lampl, Michelle, Amanda Mummert, and Meriah Schoen. 2016. "Promoting Healthy Growth or Feeding Obesity?" *Healthcare* 4(84). doi:10.3390 http://www.mdpi. com/2227-9032/4/4/84

Langa, Lungi. 2010. "Breast is Always Best, Even for HIV-positive Mothers." *Bulletin of the World Health Organization.* 88(1) January. http://www.who.int/ bulletin/volumes/88/1/10-030110/en/

Lazarov, Minda, and Amy Evans. 2000. "Breastfeeding— Encouraging the Best for Low-Income Women." *Zero to Three.* August/September. 21 (1): 15-23. http://citeseerx. ist.psu.edu/viewdoc/download?doi=10.1.1.197.8189&rep =rep1&type=pdf

Lee, Robyn. 2016. "Feeding the Hungry Other: Levinas, Breastfeeding, and the Politics of Hunger." *Hypatia.* Spring. 31 (2): 259-274. http://onlinelibrary.wiley.com/ doi/10.1111/hypa.2016.31.issue-2/issuetoc

Lee, Theodore T., Abbe R. Gluck, and Gregory Curfman. 2016. "The Politics of Medicare and Drug-Price Negotiation." *Health Affairs Blog.* September 19. http://healthaffairs. org/blog/2016/09/19/the-politics-of-medicare-and-drug-price-negotiation/

Levine, Jessica. 2016. "Chinese Mothers Bank on Breast Milk." *Sixth Tone.* October 20. http://www.sixthtone.com/news/ chinese-mothers-bank-breast-milk#

Life in Saudi Arabia. 2016. "No Infant to Sleep Hungry in the Kingdom—Government to Provide Free Baby Milk." Life in Saudi Arabia. http://life-in-saudiarabia.blogspot. com.eg/2016/11/no-infant-to-sleep-hungry-in-kingdom. html#.WDHsx04B6QY.facebook

Lisa. 2015. *The Continuum of Growth in Global Demand for Infant Formula Milk*. Lisa Global Nutrition Expertise. http://www.lisa-nutrition.com/technical--economic-environment-of-infant-milk-powder#global-demand-growth

Lutkin, Aimée. "A Mom's Side by Side Picture of Breast Milk is Blowing Everyone's Mind." *Distractify*. February 23. http://distractify.com/trending/2016/02/23/aimee-gross-photo-of-breast-milk-is-fascinating-the-internet

Mahfouz, Heba. 2016. "To get subsidized baby formula, Egyptian women need breast examinations." *Washington Post*. September 21. https://www.washingtonpost.com/news/worldviews/wp/2016/09/21/to-get-subsidized-baby-formula-egyptian-women-need-a-breast-examination/

Margulies, Leah. 1997. "The International Code of Marketing of Breastmilk Substitutes: A Model for Assuring Children's Nutrition Rights Under the Law." *International Journal of Children's Rights*. 5 (4): 419-438. http://www2.hawaii.edu/~kent/margulies.html

Mathematica Policy Research. 2015. *WIC Breastfeeding Policy Inventory: Final Report*. Submitted to United States Department of Agriculture. January. http://www.fns.usda.gov/sites/default/files/ops/WICBPI.pdf

medGadget. 2016 New *Research on Global Infant Formula Testing Industry 2016, Trends and Forecast Report*. medGadget. December 6. http://www.medgadget.com/2016/12/new-research-on-global-infant-formula-testing-industry-2016-trends-and-forecast-report.html

Medical Xpress. 2015. *Freeze-drying Breast Milk Retains More of its Healthy Properties.* December 17. http://medicalxpress.com/news/2015-12-freeze-drying-breast-retains-healthy-properties.html

Medolac Laboratories. 2015. "Medolac® Laboratories Launches Donormilk.com, the First Ever Direct-to-Consumer Offering of Human Donor Milk." *Yahoo! Finance.* May 4. http://finance.yahoo.com/news/medolac-laboratories-launches-donormilk-com-135700374.html

Metro Briefs. 2016. "Mandaluyong Moms Help Keep Milk Flowing for PGH." *Philippine Daily Inquirer.* November 19. http://newsinfo.inquirer.net/845776/metro-briefs-6

Memoria Chilena 2016. *Gotas de leche (1900-1940).* http://www.memoriachilena.cl/602/w3-article-100643.html

Middle East Monitor. 2016. "Egypt Cuts Baby Formula Subsidies in Latest Attempt to Save Economy." *Middle East Monitor.* September 2. https://www.middleeastmonitor.com/20160902-egypt-cuts-baby-formula-subsidies-in-latest-attempt-to-save-economy/

Milk Powder Solutions Pty Ltd—Buenos Aires Office. 2016. Website. http://www.21food.com/showroom/75488/product/infant-formula.html

Minchin, Maureen. 1998a. *Breastfeeding Matters: What We Need to Know About Infant Feeding.* Fourth Revised Edition. St. Kilda, Australia: Alma Publications.

---. 1998b. *Artificial Feeding: Risky for Any Baby.* St. Kilda, Australia: Alma Publications.

---. 2015a. "Excerpt from *Milk Matters: Infant Feeding & Immune Disorder.*" http://leadertoday.breastfeedingtoday-llli.org/milk-matters-infant-feeding-and-immune-disorder/

---. 2015b. *Milk Matters: Infant Feeding & Immune Disorder*. Milk Matters. Pty Ltd.

---. 2016. *Infant Formula and Modern Epidemics: The Milk Hypothesis*. Kindle Edition. https://www.amazon.com/ Infant-Formula-Modern-Epidemics-hypothesis-ebook/ dp/B01M1NJOYR/ref=sr_1_4?ie=UTF8&qid=1477155 463&sr=8-4&keywords=minchin

Ministerio de Salud. 2011. *Manual de Programas Alimentarios*. Santiago, Chile: Ministerio de Salud, Subsecretaría de Salud Pública. http://web.minsal.cl/portal/url/item/ caa1783ed97a1425e0400101640109f9.pdf

---. 2013. *Informe Técnico: Encuesta Nacional de Lactancia Materna en la Atención Primaria – (ENALMA)*. Chile: Ministerio de Salud, Subsecretaría de Salud Pública. http://web. minsal.cl/sites/default/files/INFORME_FINAL_ ENALMA_2013.pdf

---. 2014. *Norma Técnica para la supervisión de niños y niñas de 0 a 9 años en la Atención Primaria de Salud*. Chile: Ministerio de Salud, Subsecretaría de Salud Pública. http://web.minsal.cl/sites/default/files/files/2014_ Norma%20T%C3%A9cnica%20para%20la%20 supervisi%C3%B3n%20de%20ni%C3%B1os%20y%20 ni%C3%B1as%20de%200%20a%209%20en%20APS_ web(1).pdf

---. 2016a. *Proyecto de incorporación de Fórmula de Inicio en el Programa Nacional de Alimentación Complementaria (PNAC). Propuesta de implementación*. Santiago, Chile: Ministerio de Salud, Subsecretaría de Salud Pública. http://ciperchile.cl/wp-content/uploads/Proyecto-incorporacioon-FI-PNAC.pdf

---. 2016b. *Protocolo de incorporación de Fórmula de Inicio en el Programa Nacional de Alimentación Complementaria (PNAC). Propuesta de implementación.* Santiago, Chile: Ministerio de Salud, Subsecretaría de Salud Pública.

---. 2016c. *Proyecto piloto de incorporación de fórmula de inicio en el PNAC.* Santiago, Chile: Ministerio de Salud, Subsecretaría de Salud Pública. http://web.minsal. cl/wp-content/uploads/2015/10/Proyecto-Piloto-F%C3%B3rmula-de-Inicio-en-PNAC.pdf

---. 2016d. *Minuta aclaratoria: Proyecto piloto de incorporación de fórmula de inicio en el PNAC.* Santiago, Chile: Ministerio de Salud, Subsecretaría de Salud Pública. July 14.

Mintz, Morton. 2012. *The Dirty War vs. Mother's Milk.* Unpublished manuscript. https:// thedirtywarvsmothersmilk.wordpress.com/about-the-author/

Miris. 2015. *Human Milk Analysis Products and Accessories.* Miris Corporation. http://miris.se/human-milk-analysis/ products/

Monteiro, Carlos 2011. "The big issue is ultra-processing. 'Carbs': The answer." *World Nutrition*, February. 2 (2): 86-97 Accessible through *Health Impact News*, at http://www. wphna.org/htdocs/2011_feb_wn4_cam5.htm

Monteiro, Carlos A., Geoffrey Cannon, Renata Levy, Jean-Claude Moubarac, Patricia Jaime, Ana Paula Martins, Daniela Canella, Maria Louzada, Diana Parra. Also with Camila Ricardo, Giovanna Calixto, Priscila Machado, Carla Martins, Euridice Martinez, Larissa Baraldi, Josefa Garzillo, Isabela Sattamini. 2016. "Nova. The Star Shines Bright." *World Nutrition*. January-March. 7 (1-3). 28-38. http://wphna.org/wp-content/uploads/2016/01/WN-2016-7-1-3-28-38-Monteiro-Cannon-Levy-et-al-NOVA.pdf

Mooney, Phil. 2008. *Coke and the U.S. Troops*. Coca-Cola Journey. November 11. http://www.coca-colacompany.com/stories/coke-and-the-us

Moss, Michael. 2013. *Salt, Sugar, Fat: How the Food Giants Hooked Us*. New York: Random House.

Mothers Milk Cooperative 2016. Website. http://www.mothersmilk.coop/

National Alliance for Breastfeeding Advocacy. 2016. *NABA Real: Code Monitoring*. NABA. http://www.naba-breastfeeding.org/nabareal.htm

National Institutes of Health. 2016. *HIV therapy for breastfeeding mothers can virtually eliminate transmission to babies.* July 18. Washington, D.C.: NIH. https://www.nih.gov/news-events/news-releases/hiv-therapy-breastfeeding-mothers-can-virtually-eliminate-transmission-babies

National WIC Association. 2014. "The Role of Infant Formula in the WIC Program." Washington, D.C.: NWA. https://s3.amazonaws.com/aws.upl/nwica.org/role-of-infant-formula-in-wica.pdf

---. 2016. *What the Dissolution of Relationships with Infant Formula Manufacturers Means*. Washington, D.C. NWA. September 12. https://www.nwica.org/blog/what-the-dissolution-of-relationships-with-infant-formula-manufacturers-means#.WAQMmNx9fJA

Neuberger, Zoë. 2010. *WIC Food Package Should Be Based on Science: Foods with New Functional Ingredients Should Be Provided Only if They Deliver Health or Nutritional Benefits*. Washington, D.C.: Center on Budget and Policy Priorities. http://www.cbpp.org/files/6-4-10fa.pdf

Newmark, Lauren Milligan. 2016. "From Mother's Gut to Milk." *Splash!* International Genomics Consortium. April. http://milkgenomics.org/article/from-mothers-gut-to-milk/

NICE. 2010. *Donor Milk Banks: The Operation of Donor Milk Bank Services*. National Institute for Health Care and Excellence. United Kingdom. http://www.nice.org.uk/guidance/CG93

Novac, Viveca. 1988. "Formula for Profit: How Private Corporations Grow Fat From a Program Designed to Feed the Poor." *Seeds Ending U.S. and World Hunger*. August.

Oaklander, Mandy. 2016. "Many Foods Subsidized by the Government Are Unhealthy." *Time*. July 5. http://time.com/4393109/food-subsidies-obesity/

OECD 2016. *OECD Family Database*. Paris: Organisation for Economic Co-operation and Development. http://www.oecd.org/els/family/database.htm

OHCHR. 1990. *Convention on the Rights of the Child*. Geneva: Office of the High Commissioner for Human Rights. http://ohchr.org/EN/ProfessionalInterest/Pages/CRC.aspx

---. 2016. *Ratification Status of Human Rights Instruments.* United Nations: Office of the High Commissioner for Human Rights. http://indicators.ohchr.org/

Oliveira, Victor and Elizabeth Frazão. 2015a. "Painting a More Complete Picture of WIC: How WIC Impacts Nonparticipants." *Amber Waves.* April 6. http://www. ers.usda.gov/amber-waves/2015-april/painting-a-more-complete-picture-of-wic-how-wic-impacts-nonparticipants.aspx#.VSz97VyeO61

Oliveira, Victor and Elizabeth Frazão. 2015b. *The WIC Program: Background, Trends, and Economic Issues, 2015 Edition.* Washington, D.C.: United States Department of Agriculture. Economic Information Bulletin Number 134. http://www.ers.usda.gov/publications/pub-details/?pubid=43927

Oliveira, Victor, Elizabeth Frazão, and David Smallwood. 2010. *Rising Infant Formula Costs to the WIC Program: Recent Trends in Rebates and Wholesale Prices.* Washington, D.C.: U.S. Department of Agriculture. Economic Research Report Number 93. http://www.ers.usda.gov/media/136568/err93_1_.pdf

Oliveira, Victor, Elizabeth Racine, Jennifer Olmsted, and Linda M. Ghelfi. 2002. *The WIC Program: Background, Trends, and Issues.* Washington, D.C: United States Department of Agriculture. Food Assistance and Nutrition Research Report Number 27.

Only the Breast 2016. Website. http://www.onlythebreast.com/

Oshaug, Arne, and Grete Synøve Botten. 1993. "Human Milk in Food Supply Statistics." *Food Policy.* 19 (5) 479-482.

Palmer, Gabrielle 2009. *The Politics of Breastfeeding: When Breasts are Bad for Business*. London: Pinter & Martin.

Palmquist, Aunchalee E. L., and Kirsten Doehler. 2014. "Contextualizing Online Human Milk Sharing: Structural Factors and Lactation Disparity Among Middle Income Women in the U.S." *Social Science & Medicine*. 122: 140-147.

Pan, Liping, David S. Freedman, Andrea J. Sharma, Karen Castellanos-Brown, Sohyun Park, Ray B. Smith, and Heidi M Blanck. 2016."Trends in Obesity Among Participants Aged 2–4 Years in the Special Supplemental Nutrition Program for Women, Infants, and Children — United States, 2000–2014. *Morbidity and Mortality Weekly Report*. November 18. 65 (45): 1256-1260. https://www.cdc.gov/mmwr/volumes/65/wr/mm6545a2.htm?s_cid=mm6545a2_x

Patel, Atish 2014. "India's growing breast milk banking network." *News India*. July 16. http://www.bbc.com/news/world-asia-india-28106559

PATH. 2013. *Strengthening Human Milk Banking: A Global Implementation Framework, Version 1.1*. Seattle, Washington: Bill & Melinda Gates Foundation Grand Challenges Initiative. https://www.path.org/publications/files/MCHN_strengthen_hmb_frame_Jan2016.pdf

---. 2015. *A Donation from the Heart: Human Milk Banks May Help Save Vulnerable Babies*. http://www.path.org/projects/milk-banks.php

Patlan, Kelly Lawrence, and Michele Mendelson. 2016. "WIC Participant and Program Characteristics 2014: Food Package Report," *Insight Policy Research for the U.S. Department of Agriculture.* February. http://www.fns.usda. gov/sites/default/files/ops/WICPCFoodPackage2014.pdf

Payne, Sarah and Maria A. Quigley. 2016. "Breastfeeding and Infant Hospitalization: Analysis of the UK 2010 Infant Feeding Survey." *Maternal & Child Nutrition.* March 24. http://onlinelibrary.wiley.com/doi/10.1111/mcn.12263/ abstract

Petherick, Anna. 2016. "Breast Milk, the Synchronizer." *Splash!* Milk Science Update. International Milk Genomics Consortium. February. http://milkgenomics.org/article/ milk-the-synchronizer/?utm_sour%E2%80%A6letter_ February2016&utm_campaign=SPLASHfeb2016&utm_ medium=email

Physicians Committee for Responsible Medicine. 2016. *Government Support for Unhealthful Foods.* PCRM blog. http://www.pcrm.org/health/reports/agriculture-and-health-policies-unhealthful-foods

Piwoz, Ellen G. and Sandra L. Huffman 2015. "The Impact of Marketing of Breast-Milk Substitutes on WHO-Recommended Breastfeeding Practices." *Food and Nutrition Bulletin.* August 27. http://fnb.sagepub.com/ content/early/2015/08/26/0379572115602174.full. pdf+html

PRWEB. 2014. *Human Milk Banking Association of North America Takes a Stand Against Paying for Donations.* PRWeb. December 9. http://www.prweb.com/ releases/2014/12/prweb12374945.htm

Raju, Tonse N. K. 2011. "Breastfeeding is a Dynamic Biological Process—Not Simply a Meal at the Breast." *Breastfeeding Medicine.* October. 6 (5): 257-259. https://www.ncbi.nlm.nih.gov/pmc/articles/PMC3199546/

Ralston, Katherine L. 2006. *Nutrient Adequacy of Children Participating in WIC.* United States Department of Agriculture. Economic Research Service. Economic Brief Number 8. April. http://purl.umn.edu/34091

Rama de Nutrición. 1999. "Leche Purita Fortificada en la alimentación infantil." *Revista Chilena de Pediatría,* July. http://www.scielo.cl/scielo.php?script=sci_arttext&pid=S0370-41061999000400018

Richter, Judith. 2001. *Holding Corporations Accountable: Corporate Conduct, International Codes and Citizen Action.* London: Zed Books.

Rinker, Brian. 2016. *New Law Aims to Regulate For-Profit Human Breast Milk Banks.* California Healthline. April 6. http://californiahealthline.org/news/new-law-aims-to-regulate-for-profit-human-breast-milk-banks/

Rohde, Jon Eliot 1982. "Mother Milk and the Indonesian Economy: A Major National Resource." *Journal of Tropical Pediatrics.* (4): 166-174.

Rollins, Nigel C., Nita Bhandari, Neemat Hajeebhoy, Susan Horton Chessa K. Lutter, Jose C. Martines, Ellen G. Piwoz, Linda M. Richter, and Cesar G. Victora, on behalf of *The Lancet* Breastfeeding Series Group. 2016"Why Invest, and What it Will Take to Improve Breastfeeding Practices." *The Lancet,* January. 387 (10017). http://www.thelancet.com/pdfs/journals/lancet/PIIS0140-6736(15)01044-2.pdf

Ryan, Alan S. and Wenjun Zhou. 2006. "Lower Breastfeeding Rates Persist Among the Special Supplemental Nutrition Program for Women, Infants, and Children Participants, 1978-2003." *Pediatrics*. April. http://pediatrics. aappublications.org/content/117/4/1136

Sadeq-Nevine, Georgette, and Kameel-Claire Sidqy. 2016. "Subsidised Baby Milk: Is There a Shortage?" *Watani International*. September 21. http://en.wataninet. com/features/health/subsidised-baby-milk-is-there-a-shortage/17422/

Saitone, Tina L., Richard J. Sexton, and Richard J. Volpe, III 2015. "A Wicked Problem? Cost Containment in the Women, Infants and Children Program." *Applied Economic Perspectives and Policy*. 37 (3): 378-402. http:// aepp.oxfordjournals.org/content/37/3/378.abstract

Salcedo, Jamie; Maria Gormaz; Maria C. López-Mendoza; Elisabetta Nogaroot; Dolores Silvestre 2015. "Human Milk Bactericidal Properties: Effect of Lyophilization and Relation to Maternal Factors and Milk Components." *Journal of Pediatric Gastroenterology and Nutrition*. April. 60 (4): 527-532. http://journals.lww.com/jpgn/pages/ articleviewer.aspx?year=2015&issue=04000&article=0002 3&type=abstract

Saleh, Heba. 2016. "President Sisi Deploys Army to Tackle Egypt's Economic Woes." *Financial Times*. October 4. https://www.ft.com/content/00ea1c04-8a14-11e6-8cb7-e7ada1d123b1app?action=DocumentDisplay&crawlid=1 &doctype=cite&docid=2+J.+Pharmacy+%26+Law+185& srctype=smi&srcid=3B15&key=fb6a4d617b7fbb66e342d e3bef1bba57

Sankar, Mari Jeeva, Bieshwar Sinha, Ranadip Chowdhury, Nita Bhandari, Sunta Taneja, Jose Martines, and Rajiv Bahl. 2015. "Optimal Breastfeeding Practices and Infant and Child Mortality: A Systematic Review and Meta-analysis." *Acta Pædiatrica.* 104 (S467): 3-13. http://onlinelibrary.wiley.com/enhanced/doi/10.1111/apa.13147

Schaal, Benoist, Gérard Coureaud, Dominique Langlois, Christan Giniès. Etienne Sémon, and Guy Perrier. 2003. "Chemical and Behavioural Characterization of the Rabbit Mammary Pheromone." *Nature.* 424: 68-72. doi:10.1038/nature01739

Schiller, Ben. 2016. "It's Time for A Global System of Breast Milk Banks." *Co.Exist Daily.* September 28. https://www.fastcoexist.com/3063989/its-time-for-a-global-system-of-breast-milk-banks

Shubber, Sami 1985. *The International Code, Digest of Health Legislation.* 36 (4). http://ibfan.org/what-is-the-international-code

---. 1998. *The International Code of Marketing of Breast-Milk Substitutes.* The Hague: Kluwer.

---. 2011. *The WHO International Code of Marketing of Breast-Milk Substitutes: History and Analysis.* Second edition. London, U.K.: Pinter & Martin.

Smith, Julie P. 2013. "'Lost Milk?': Counting the Economic Value of Breast Milk in Gross Domestic Product." *Journal of Human Lactation.* 29 (4): 537-546. https://www.researchgate.net/publication/249322424_Lost_Milk_Counting_the_Economic_Value_of_Breast_Milk_in_Gross_Domestic_Product

---. 2015. "Markets, Breastfeeding and Trade in Mother's Milk." *International Breastfeeding Journal*, 10 (9). http://internationalbreastfeedingjournal.biomedcentral.com/articles/10.1186/s13006-015-0034-9

Smith, Julie; Judith Galtry, and Libby Salmon. 2014. "Confronting the Formula Feeding Epidemic in a New Era of Trade and Investment Liberalisation." *Journal of Australian Political Economy*. 73. http://media.wix.com/ugd/b629ee_95b1495d485d47e280a5b74d64e70cf0.pdf

Smith, Julie, Libby Salmon, and Phillip Baker. 2016. "World Breastfeeding Week: Conflicts of Interest in Infant and Young Child Feeding." *PLOS Blogs: Translational Global Health*. http://blogs.plos.org/globalhealth/2016/08/world-breastfeeding-week-conflicts-of-interest-in-infant-and-young-child-feeding/

Smith, Lawrence Weld. 1924. *The Experimental Feeding of Dried Breast Milk*. June 17. http://www.jbc.org/content/61/3/625.full.pdf

Springwise. 2016. *Milk Bank Pays Cambodian Mothers for Their Excess Breast Milk*. Springwise. March 23. https://www.springwise.com/milk-bank-pays-cambodian-mothers-excess-breast-milk/

Stevens, Emily E., Thelma E. Patrick, and Rita Pickler. 2009. "A History of Infant Feeding." *Journal of Perinatal Education*. 18 (2): 32-39. https://www.ncbi.nlm.nih.gov/pmc/articles/PMC2684040/pdf/jpe-18-032.pdf

Stipicic, Mónica. 2016. "The Milk Battle." *La Tercera*. August 13. http://www.latercera.com/noticia/tendencias/2016/08/27-692307-9-la-batalla-de-la-leche.shtml#last

Strader, Kristen. 2016. *Infant Formula Marketing in Public Hospitals: An Outdated and Ethical Practice.* Washington, D.C.: Public Citizen. April. http://www.citizen.org/documents/public-hospitals-infant-formula-marketing-report-april-2016.pdf

Strom, Stephanie. 2016. "How a Food Subsidy Program Pushed Junk Food Off the Table." *New York Times.* October 6. http://www.nytimes.com/2016/10/05/well/eat/how-a-food-subsidy-program-pushed-junk-food-off-the-table.html?_r=0

Swanson, Kara W. 2014. *Banking on the Body: The Market in Blood, Milk, and Sperm in Modern America.* Cambridge, Massachusetts: Harvard University Press.

The Alpha Parent. 2015. "Is Formula Feeding Worse Than Smoking?" July 7. *The Politics of Parenting Blog.* http://www.thealphaparent.com/2014/07/is-formula-feeding-worse-than-smoking.html

The Economist 2006. "The baby-food king." *The Economist.* September 2. http://www.economist.com/node/7854488

The Lancet. 2016. *Breastfeeding.* January 29. http://www.thelancet.com/series/breastfeeding

The Milk Meg. 2013. *Ingredients in Breastmilk Versus Artificial Breastmilk (Formula).* http://themilkmeg.com/ingredients-in-breastmilk-versus-artificial-breastmilk-formula/

Thorley, Virginia 2015. "Free Supplies and the Appearance of Endorsement: Distribution of Tobacco to Soldiers and Artificial Infant Formula to New Mothers." *Journal of Human Lactation.* 31(2): 213-215. https://www.researchgate.net/publication/271331195_Free_Supplies_and_the_Appearance_of_Endorsement_Distribution_of_Tobacco_to_Soldiers_and_Artificial_Infant_Formula_to_New_Mothers

Thorley, Virginia. 2016. "Milk Kinship and Implications for Human Milk Banking: A Review." *Women's Health Bulletin.* 3 (3). e36897. http://womenshealthbulletin.com/?page=article&article_id=36897

Thurow, Roger. 2016. *The First 1,000 Days: A Crucial Time for Mothers and Children.* Chicago Public Affairs.

Toumi, Habib. 2016. "Free Milk for Needy Babies in Saudi Arabia." *Gulf News: Saudi Arabia.* May 17. http://gulfnews.com/news/gulf/saudi-arabia/free-milk-for-needy-babies-in-saudi-arabia-1.1829150

Tuttle, Cynthia Reeves. 2000. "An Open Letter to the WIC Program: The Time Has Come to Commit to Breastfeeding." *Journal of Human Lactation.* May. 16 (2): 99-103.

UNICEF 2013. *Breast Milk Banks are a Sound Investment in the Health of Brazil's Premature Babies.* New York: United Nations Children's Fund. http://www.unicef.org/infobycountry/brazil_70944.html

---. 2016a. *Adopting Optimal Feeding Practices is Fundamental to a Child's Survival, Growth, and Development, But Too Few Children Benefit.* New York: United Nations Children's Fund. http://data.unicef.org/topic/nutrition/infant-and-young-child-feeding/#

---. 2016b. Press Release: *77 Million Newborns Globally Not Breastfed Within First Hour of Life.* New York: United Nations Children's Fund. http://www.unicef.org/media/media_92038.html

---. 2016 c. *The Baby-Friendly Hospital Initiative.* New York: United Nations Children's Fund. http://www.unicef.org/programme/breastfeeding/baby.htm

---. 2016d. *International Code of Marketing of Breast-milk Substitutes.* New York: United Nations Children's Fund. http://www.unicef.org/nutrition/index_24805.html

---. 2016e. *From the First Hour of Life: A New Report on Infant and Young Child Feeding.* New York: United Nations Children's Fund. https://data.unicef.org/resources/first-hour-life-new-report-breastfeeding-practices/

UNICEF and WHO, 2015. *Breastfeeding Advocacy Initiative: For the best start in life.* New York and Geneva: *United Nations* Children's Fund and World Health Organization. http://www.unicef.org/nutrition/files/Breastfeeding_Advocacy_Strategy-2015.pdf

Union of Concerned Scientists. 2016. *Unhealthy Food Policy.* UCS blog. http://www.ucsusa.org/our-work/food-agriculture/our-failing-food-system/unhealthy-food-policy#.WCUsOtx9fJB

United Nations. Economic and Social Council. Committee on Economic, Social and Cultural Rights. 1999. *General Comment 12: Substantive Issues Arising in the Implementation of the International Covenant on Economic, Social and Cultural Rights: General Comment 12 (Twentieth Session, 1999) The Right to Adequate Food (art. 11)* (Geneva: ECOSOC E/C.12/1999/5). http://www.fao.org/fileadmin/templates/righttofood/documents/RTF_publications/EN/General_Comment_12_EN.pdf

---. 2000. *General Comment 14: United Nations. Economic and Social Council. Committee on Economic, Social and Cultural Rights. The Right to the Highest Attainable Standard of Health (Article 12 of the International Covenant on Economic, Social and Cultural Rights)* (Geneva: ECOSOC E/C.12/2000/4). http://www.ohchr.org/Documents/Issues/Women/WRGS/Health/GC14.pdf

United Nations. General Assembly. Human Rights Council. 2016. *The Right to Food.* March 21. Geneva: United Nations. A/HRC/31/L.14. https://documents-dds-ny.un.org/doc/UNDOC/LTD/G16/056/85/PDF/G1605685.pdf?OpenElement

United States Breastfeeding Committee 2016. Website. http://www.usbreastfeeding.org/

United States Centers for Disease Control and Prevention. 2016a. *Breastfeeding.* Atlanta, Georgia: CDC. http://www.cdc.gov/breastfeeding/

---. 2016b. *Breastfeeding. National Policies and Positions.* Atlanta, Georgia: CDC. http://www.cdc.gov/breastfeeding/policy/index.htm

---. 2016c. *Breastfeeding. Promotion and Support.* Atlanta, Georgia: CDC.: http://www.cdc.gov/breastfeeding/promotion/index.htm

United States Department of Agriculture. 2004. "Study examines long-term health effects of soy infant formula." *Agricultural Research.* January. 52 (2): 8-10. http://www.ars.usda.gov/is/AR/archive/jan04/soy0104.pdf

---. 2016a. *Loving Support: Makes Breastfeeding Work: FY 2015.* Washington, D.C. USDA. http://www.fns.usda.gov/sites/default/files/wic/FY%202015%20BFDLA%20Report.pdf

---. 2016b. *WIC Program Participation and Costs (Data as of October 7, 2016).* Washington, D.C.: USDA. http://www.fns.usda.gov/sites/default/files/pd/wisummary.pdf

---. 2016c. *Women, Infants and Children (WIC): About WIC-WIC's Mission.* http://www.fns.usda.gov/wic/about-wic-wics-mission

---. 2016d. *Women, Infants and Children (WIC): Frequently Asked Questions About WIC.* Washington, D.C.: USDA. http://www.fns.usda.gov/wic/frequently-asked-questions-about-wic

---. 2016e. *Women, Infants and Children (WIC): Breastfeeding Promotion in WIC: Current Federal Requirements.* Washington, D.C.: USDA. http://www.fns.usda.gov/wic/breastfeeding-promotion-wic-current-federal-requirements

---. 2016f. *Women, Infants and Children (WIC). Breastfeeding Promotion and Support.* Washington, D.C.: USDA. http://www.fns.usda.gov/wic/breastfeeding-promotion-and-support-wic

---. 2016g. *Women, Infants and Children (WIC): WIC Laws and Regulations.* Washington, D.C.: USDA. http://www.fns.usda.gov/wic/wic-laws-and-regulations

---. 2016h. *Women, Infants and Children (WIC): About WIC—How WIC Helps.* Washington, D.C.: USDA. http://www.fns.usda.gov/wic/about-wic-how-wic-helps#Improved%20Diet%20and%20Diet-Related%20Outcomes

---. 2016i. *Egypt: Government Reforms its Infant Milk Formula Program Effective August 1, 2016, and Sources U.S. Infant Milk Formula to Address Market Shortages.* USDA Foreign Agricultural Service: Global Agricultural Information Network. July 26. http://gain.fas.usda.gov/Recent%20GAIN%20Publications/Government%20Reforms%20its%20Infant%20Milk%20Formula%20Program%20Effective%20August%20_Cairo_Egypt_7-26-2016.pdf

United States Department of Health and Human Services. 2011. *The Surgeon General's Call to Action to Support Breastfeeding.* Washington, D. C.: Office of the Surgeon General. http://www.surgeongeneral.gov/topics/breastfeeding/index.html

United States Food and Drug Administration. 2015. *Use of Donor Human Milk.* Washington, D.C.: U.S. Department of Health & Human Services. Food and Drug Administration. http://www.fda.gov/ScienceResearch/SpecialTopics/PediatricTherapeuticsResearch/ucm235203.htm

---. 2016. *FDA Issues Draft Guidance Concerning the Type and Quality of Scientific Evidence Underlying Structure/Function Claims Made in Infant Formula Labels and Labeling.* Washington, D.C.: U.S. Department of Health & Human Services. Food and Drug Administration. http://www.fda.gov/Food/NewsEvents/ConstituentUpdates/ucm514389.htm

United States Government Accountability Office. 2006. *Breastfeeding: Some Strategies Used to Market Infant Formula May Discourage Breastfeeding; State Contracts Should Protect Against Misuse of WIC Name.* USGAO. GAO-06-282. http://www.gao.gov/new.items/d06282.pdf

University of California—Riverside. 2016. *Vaccinating Babies Without Vaccinating Babies: A Baby Makes Copies of Maternal Immune Cells it Acquires Through Mother's Milk.* Science News. https://www.sciencedaily.com/releases/2016/10/161007085631.htm?utm_source=feedburner&utm_medium=email&utm_campaign=Feed%3A+sciencedaily%2Fhealth_medicine%2Fbreastfeeding+%28Breastfeeding+News+--+ScienceDaily%29

Vigliarolo, Marie and Veronica Diaz. 2016. *Quick International Delivers Shipment of Precious Breast Milk to Orphanage in South Africa for Premature, Sick and Orphaned Infants.* PRWEB. November 15. http://www.prweb.com/releases/2016/11/prweb13850734.htm

Viñals, Victoria. 2015. *Abren Primer Banco de Leche para Prematuros Extremos.* Diario Uchile. September 15. http://radio.uchile.cl/2015/09/20/abren-primer-banco-de-leche-humana-en-chile-para-prematuros-extremos/

Vittachi, Anuradha. 1982. "Stop the Babymilk Pushers." *New Internationalist*. April 1. https://newint.org/features/1982/04/01/keynote/

Wardlaw, Tessa, Danzhen You, Lucia Hug, Agbessi Amouzou, and Holly Newby. 2014."UNICEF Report: Enormous Progress in Child Survival but Greater Focus on Newborns Urgently Needed." *Reproductive Health*. 11 (82). http://data.unicef.org/corecode/uploads/document6/uploaded_pdfs/corecode/Enormous-progress-in-child-survival_220.pdf

Wattana, Melissa. 2016. *The Baby Bottle and the Bottom Line: Corporate Strategies and the Infant Formula Controversy in the 1970s*. Yale University. Senior Essay. http://hshm.yale.edu/sites/default/files/files/Wattana%20senior%20essay%202016.pdf

Winberg, Jan. 2005. "Mother and Newborn Baby: Mutual Regulation of Physiology and Behavior—A Selective Review." *Developmental Psychobiology*. November. 47 (3): 217-229. DOI: 10.1002/dev.20094 http://onlinelibrary.wiley.com/doi/10.1002/dev.20094/abstract;jsessionid=3611DE5AF12B6195A189CF4DA669DD63.f02t01

Wirtschafter, Jacob. 2016. "Egypt Requires Breast Exam for Subsidized Formula." The Medialine. September 26. http://www.themedialine.org/top-stories/egypt-requires-breast-exam-subsidized-formula/

Wood, Benjamin. 2015. "Got breast milk? If Not, a Utah County Company Has You Covered with Imported Cambodian Milk." *The Salt Lake Tribune*. December 22. http://www.sltrib.com/news/3340606-155/got-breast-milk-if-not-a

Wood, Laura. 2011. "The Welfare State and Mother's Milk." *The Thinking Housewife*. August 10. http://www. thinkinghousewife.com/wp/2011/08/the-welfare-state-and-mothers-milk/

World Bank. 2015. *World Development Indicators: Fragile Situations Part 1*. World Bank. http://wdi.worldbank.org/table/5.8

World Health Assembly. 1986. Infant and Young Child Feeding. http://www.who.int/nutrition/topics/WHA39.28_iycn_en.pdf?ua=1

---. 1994. Resolution WHA47.5 Infant and Young Child Nutrition. http://www.who.int/nutrition/topics/WHA47.5_iycn_en.pdf?ua=1

---. 2016a. *Maternal, Infant and Young Child Nutrition: Ending Inappropriate Promotion of Foods for Infants and Young Children*. Sixty-Ninth World Health Assembly. Geneva: World Health Organization. May 28. WHA 69.9. http://apps.who.int/gb/ebwha/pdf_files/WHA69/A69_R9-en.pdf

---. 2016b. Guidance on Ending the Inappropriate Promotion of Foods for Infants and Young Children. A69/7 Add.1 May 13, 2016. http://apps.who.int/gb/ebwha/pdf_files/WHA69/A69_7Add1-en.pdf?ua=1

World Health Organization. 1981. *International Code of Marketing of Breast-milk Substitutes*. Geneva: WHO. http://www.who.int/nutrition/publications/code_english.pdf

--- 2003. *Global Strategy for Infant and Young Child Feeding*. Geneva: WHO. http://www.who.int/nutrition/publications/infantfeeding/9241562218/en/index.html

---. 2007. *Infant and Young Child Feeding in Emergencies: Operational Guidance for Emergency Relief Staff and Programme Managers.* Geneva: WHO. February. http://www.unhcr.org/45f6cd022.pdf

---. 2008. *The International Code of Marketing of Breast-milk Substitutes: Frequently Asked Questions (Updated Version 2008).* Geneva: WHO. http://www.who.int/nutrition/publications/infantfeeding/9789241594295/en/and

---. 2013a. *Short-Term effects of Breastfeeding: A Systematic Review on the Benefits of Breastfeeding on Diarrhoea and Pneumonia Mortality.* Geneva: WHO. http://www.who.int/maternal_child_adolescent/documents/breastfeeding_short_term_effects/en/

---. 2013b. *Long-Term Effects of Breastfeeding: A Systematic Review.* Geneva: WHO. http://www.who.int/maternal_child_adolescent/documents/breastfeeding_long_term_effects/en/

---. 2013c. *Information concerning the use and marketing of follow-up formula.* July 17. Geneva: WHO. http://www.who.int/nutrition/topics/WHO_brief_fufandcode_post_17July.pdf

---. 2014. *Children: Reducing Mortality.* Geneva: WHO. http://www.who.int/mediacentre/factsheets/fs178/en/#

---. 2015. *Country Implementation of the International Code of Marketing of Breast-milk Substitutes: Status Report 2011.* Geneva: WHO. http://www.who.int/nutrition/publications/infantfeeding/statusreport2011/en/

---. 2016. *Nutrition: World Health Assembly Resolutions and Documents. Infant and Young Child Nutrition.* Geneva: WHO. http://www.who.int/nutrition/topics/wha_nutrition_iycn/en/

World Health Organization and UNICEF. 2016. *Guidelines: Updates on HIV and Infant Feeding.* WHO and UNICEF. Geneva: WHO. http://apps.who.int/iris/bitstre am/10665/246260/1/9789241549707-eng.pdf

World Health Organization, UNICEF, and IBFAN. 2016. *Marketing of Breast-milk Substitutes: National Implementation of the International Code: Status Report 2016.* WHO, UNICEF, IBFAN. Geneva: WHO. http:// www.who.int/nutrition/publications/infantfeeding/code_report2016/en/

Yunus, Muhammad. 2007. *Creating a World Without Poverty: Social Business and the Future of Capitalism.* New York: Public Affairs.

Zimmerman, Rachel 2016. "Study: Breastfeeding Even More of a Health issue for Moms Than for Babies." *CommonHealth.* September 29. http://www.wbur.org/ commonhealth/2016/09/29/study-breastfeeding-moms-health

Ziol-Guest, Kathleen M., and Daphne C. Hernandez. 2010. "First- and Second-Trimester WIC Participation is Associated with Lower Rates of Breastfeeding and Early Introduction of Cow's Milk During Infancy." *Journal of the American Dietetic Association.* May. 110 (5): 702-9. http://www.andjrnl.org/article/S0002-8223(10)00114-8/abstract

Praise for *Governments Push Infant Formula*

George Kent's book is an articulate and incisive analysis of the ways in which some governments actively promote the use of infant formula. They do this despite the predictable harm it does to children's health. The book is a timely and powerful reminder to governments in the rich and poor world of their obligations under international law to protect children's health and the right to food through framework legislation and the regulation of non-State actors including corporations. Effective remedies are urgently required.

Graham Riches
Emeritus Professor of Social Work
University of British Columbia

***

Here we have the latest advance in the breastfeeding versus formula feeding struggle. It should not be a struggle because the evidence in favor of breastfeeding has been compelling for decades. In true investigative reporting style, George Kent unlocks the door to look at whether the formula industry alone is to blame for consistently going against the evidence. He shows us that far too often, governments must carry some of the blame. Governments have failed to counter the systematic aggressive marketing strategies of the industry, and at times they have become their allies. The lingering question is whether it really amounts to collusion.

Claudio Schuftan, MD, public health nutritionist, People's Health Movement, Ho Chi Minh City.

***

Breastfeeding rates are steadily declining in all regions of the world with consequent negative effects on the health of the world's children. George Kent's new book, *Governments Push Infant Formula*, help to explain why. This is a must-read for all those in the trenches who are attempting to protect breastfeeding for the next generation.

Pamela Morrison, International Board Certified Lactation Consultant in private practice in Rustington, West Sussex, England.

***

Building on earlier work, George Kent's new book reveals how three countries actively promote artificial feeding while claiming to promote breastfeeding. Actions mean more than words. Read this book to see what actions are needed by three of the many governments that "push" infant formula. It should create many new change advocates!

Maureen Minchin, author, *Milk Matters: Infant Feeding & Immune Disorder*. See www.infantfeedingmatters.org/

***

George Kent's new book presents compelling evidence of how governments distribute large quantities of free or low-cost supplies of infant formula and other breastmilk substitutes. He documents the potential damage to infant health caused by such subsidized supplies. They undermine optimal breastfeeding practices and create a dependency on formula, the risks of which are extensively catalogued.

Kent reminds us that governments should subsidize children's health, not corporations' wealth. They should not put the business interests of corporations above the best interests of children.

Kent uses revealing analogies to underscore his argument that breastmilk, like healthy soil, is a living thing and not just a collection of parts. Breastfeeding fosters infants' optimal growth, development, and health.

Alison Linnecar, Convener, International Baby Food Action Network's Global Working Group on Chemical and Microbiological Contamination. (See www.ibfan.org/infant-and-young-child-feeding-health-and-environmental-impacts)

✳✳✳

If we now know that breastfeeding is not just best but, in fact, the human norm; if scientists now understand that human milk is not only a food but a uniquely evolved human tissue that meets the specific survival needs of the human infant; if public health experts and governments the world over now emphasize that the risks of feeding infant formula instead of breastfeeding are unequivocal; how, in good conscience, can the United States and other countries continue government programs that dispense free infant formula *en masse? Governments Push Infant Formula* is the much-needed wake-up call that it is time to face this hypocrisy head on. George Kent compellingly makes the case for a future in which widespread human milk banking would give every child the chance to thrive.

Jennifer Grayson, author of *Unlatched: The Evolution of Breastfeeding and the Making of a Controversy*

ABOUT THE AUTHOR

George Kent

After more than forty years of teaching in the University of Hawaii's Department of Political Science, George Kent retired in 2010 as Professor Emeritus. Currently he serves as an Adjunct Professor with the Department of Peace and Conflict Studies at the University of Sydney in Australia and also with the Department of Transformative Social Change at Saybrook University in California. He teaches an online course on the Human Right to Adequate Food for both of them.

Professor Kent has worked with the Food and Agriculture Organization of the United Nations, the United Nations Children's Fund, the World Food Programme and several nongovernmental organizations. He is on the Board of Directors of the International Peace Research Association Foundation. His recent books on food policy issues are *Freedom from Want: The Human Right to Adequate Food*, *Global Obligations for the Right to Food*, *Ending Hunger Worldwide*, *Regulating Infant Formula*, and *Caring About Hunger*.

www.ingramcontent.com/pod-product-compliance
Lightning Source LLC
Chambersburg PA
CBHW060551210326
41519CB00014B/3441